Two-Dimensional Constant and Product Polynomial Systems

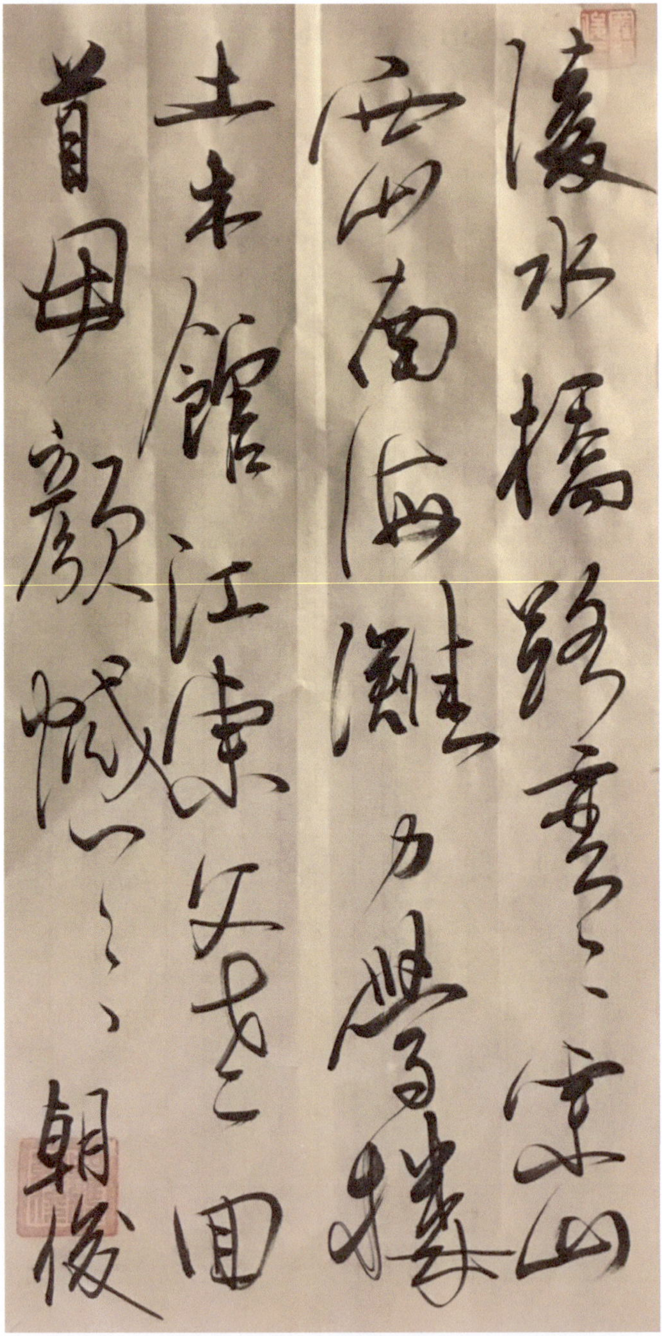

凌水桥、路弯弯，西山东山南海滩，力学楼、土木馆，江东父老、回首毋颜，憾，憾，憾。

——朝俊 回望【大连理工大学】四十年记甲辰秋

Albert C. J. Luo

Two-Dimensional Constant and Product Polynomial Systems

高等教育出版社
HIGHER EDUCATION PRESS

Springer

Albert C. J. Luo
Department of Mechanical
and Mechatronics Engineering
Southern Illinois University Edwardsville
Edwardsville, IL, USA

ISBN 978-981-96-5514-4 ISBN 978-981-96-5515-1 (eBook)
https://doi.org/10.1007/978-981-96-5515-1

Jointly published with Higher Education Press Limited Company
The print edition is not for sale in the mainland of China. Customers from the mainland of China please order the print book from: Higher Education Press Limited Company.

This Springer imprint is published by the registered company Springer Nature Singapore Pte Ltd.
The registered company address is: 152 Beach Road, #21-01/04 Gateway East, Singapore 189721, Singapore

Preface

This book is a monograph about 1-dimensional flow arrays and bifurcations in constant and product polynomial systems. The 1-dimensional flows and the corresponding bifurcation dynamics are discussed. The singular hyperbolic and hyperbolic-secant flows are presented, and the singular hyperbolic-to-hyperbolic-secant flows are discussed. The singular inflection source, sink, upper-and lower-saddle flows are presented. The corresponding appearing and switching bifurcations are presented for the hyperbolic and hyperbolic-secant networks and singular flows networks. The corresponding theorem is presented, and the proof of theorem is given. Based on the singular flows, the corresponding hyperbolic and hyperbolic-secant flows networks are illustrated for a better understanding of the dynamics of constant and product polynomial systems.

In this book, three chapters are included. Chapter 1 presents the theorem for constant and product polynomial systems. The switches between source and sink and between upper and lower saddles are presented. The dynamics of singular hyperbolic and hyperbolic-secant flows are discussed, and the singular hyperbolic-to-hyperbolic-secant flows are given. The dynamics of singular inflection-source, sink and saddle flows are also discussed. Based on the singular flows, the corresponding bifurcation dynamics are discussed. In Chap. 2, from local analysis and the first integral manifolds, the proof of theorem on the constant and product polynomial dynamics is given. In Chap. 3, the hyperbolic and hyperbolic-secant singular flows with the corresponding hyperbolic and hyperbolic-secant flow networks are presented. Based on the singular hyperbolic-to-hyperbolic-secant flows and singular hyperbolic-secant-to-hyperbolic flows, the corresponding hyperbolic and hyperbolic-secant flows are illustrated. Finally, based on the singular inflection source, sink, and saddle flows, the corresponding networks of hyperbolic and hyperbolic-secant flows are presented.

Finally, the author hopes that the materials presented herein can provide a better understanding of the 1-dimensional flows in constant and product-polynomial systems.

Edwardsville, IL, USA Albert C. J. Luo

The original version of the book has been revised. A correction to this book can be found at https://doi.org/10.1007/978-981-96-5515-1_4

Contents

Chapter 1
Constant and Product Polynomial Systems

In this chapter, nonlinear dynamics of dynamical systems possessing constant and product polynomial vector fields is discussed. There are four types of 1-dimensional singular flows: (i) The simple and singular hyperbolic and hyperbolic-secant flows with directrix flows are presented. (ii) The singular hyperbolic to hyperbolic-secant flows with directrix flows are discussed. (iii) The inflection-source and sink flows with directrix flows are presented. (iv) The inflection-upper-saddle and lower-saddle flows with directrix flows are discussed. The corresponding appearing bifurcations for the networks of hyperbolic and hyperbolic-secant flows or for the networks of singular 1-dimensional flows are from the above four types of 1-dimensional singular flows. The networks of hyperbolic and hyperbolic-secant flows are illustrated.

1.1 Constant and Product Polynomial Dynamics

In this section, the dynamics of polynomial dynamical systems with two product polynomial vector fields determined by two-different-univariate products will be presented. The corresponding dynamical behaviors will be presented through the following theorem.

Theorem 1.1 *Consider a two-dimensional, nonlinear dynamical system as*

$$\begin{aligned} \dot{x}_{j_2} &= a_{j_2 j_2 0}, \\ \dot{x}_{j_1} &= a_{j_1 j_1 0}(x_{j_1} - a_{j_1 j_1 1})^m (x_{j_2} - a_{j_1 j_2 1})^n. \end{aligned} \tag{1.1}$$

If $a_{j_2 j_2 0} = 0$, *the 2-dimensional dynamical system is a 1-dimesional system in the* x_{j_1}*-direction. The corresponding solution is*

A. C. J. Luo, *Two-dimensional Constant and Product Polynomial Systems*,
https://doi.org/10.1007/978-981-96-5515-1_1

$$
\begin{aligned}
&x_{j_2} = x_{j_2 0};\\
&-\tfrac{1}{m-1} a_{j_2 j_2 0} \Big[\frac{1}{(x_{j_1} - a_{j_1 j_1 1})^{m-1}} - \frac{1}{(x_{j_1 0} - a_{j_1 j_1 1})^{m-1}} \Big]\\
&= a_{j_1 j_1 0}(x_{j_2 0} - a_{j_1 j_2 1})^n (t - t_0),\\
&\text{for } m \neq 1;\\
&a_{j_2 j_2 0} \ln \frac{|x_{j_1} - a_{j_1 j_1 1}|}{|x_{j_1 0} - a_{j_1 j_1 1}|} = a_{j_1 j_1 0}(x_{j_2 0} - a_{j_1 j_2 1})^n (t - t_0),\\
&\text{for } m = 1.
\end{aligned}
\tag{1.2}
$$

(i_1) At $(x_{j_1}^*, x_{j_2 0}) = (a_{j_1 j_1 1}, x_{j_2 0})$ *and* $x_{j_2 0} \neq a_{j_1 j_2 1}$, *there are two cases* $m = 2m_1 + 1$ and $m = 2m_1$.

(i_{1a}) For $m = 2m_1 + 1$, *the equilibrium of* $(x_{j_1}^*, x_{j_2 0}) = (a_{j_1 j_1 1}, x_{j_2 0})$ *and* $x_{j_2 0} \neq a_{j_1 j_2 1}$ *for the* x_{j_1}*-direction has the following properties.*

- *For* $a_{j_1 j_1 0}(x_{j_2 0} - a_{j_1 j_2 1})^n > 0$ *and* $a_{j_2 j_2 0} = 0$,

$$
(a_{j_1 j_1 1}, x_{j_2 0}) = \underbrace{((2m_1+1)^{\text{th}}\text{SO,IC})}_{(2m_1+1)^{\text{th}}\text{-order source}}. \tag{1.3}
$$

The flow at $(x_{j_1}^*, x_{j_2 0}) = (a_{j_1 j_1 1}, x_{j_2 0})$ *is a* $((2m_1+1)^{\text{th}}\text{SO,IC})$*-source*

- *For* $a_{j_1 j_1 0}(x_{j_2 0} - a_{j_1 j_2 1})^n < 0$ *and* $a_{j_2 j_2 0} = 0$

$$
(a_{j_1 j_1 1}, x_{j_2 0}) = \underbrace{((2m_1+1)^{\text{th}}\text{SI,IC})}_{(2m_1+1)^{\text{th}}\text{-order sink}}. \tag{1.4}
$$

The flow at $(x_{j_1}^*, x_{j_2 0}) = (a_{j_1 j_1 1}, x_{j_2 0})$ *is a* $((2m_1+1)^{\text{th}}\text{SI,IC})$*-sink.*

(i_{1b}) For $m = 2m_1$, *the equilibrium of* $(x_{j_1}^*, x_{j_2 0}) = (a_{j_1 j_1 1}, x_{j_2 0})$ *and* $x_{j_2 0} \neq a_{j_1 j_2 1}$ *for the* x_{j_1}*-direction has the following properties.*

- *For* $a_{j_1 j_1 0}(x_{j_2 0} - a_{j_1 j_2 1})^n > 0$ *and* $a_{j_2 j_2 0} = 0$,

$$
(a_{j_1 j_1 1}, x_{j_2 0}) = \underbrace{((2m_1)^{\text{th}}\text{US,IC})}_{(2m_1)^{\text{th}}\text{-order upper-saddle}}. \tag{1.5}
$$

The flow at $(x_{j_1}^*, x_{j_2 0}) = (a_{j_1 j_1 1}, x_{j_2 0})$ *is a* $((2m_1)^{\text{th}}\text{US,IC})$*-upper-saddle*

- *For* $a_{j_1 j_1 0}(x_{j_2 0} - a_{j_1 j_2 1})^n < 0$ *and* $a_{j_2 j_2 0} = 0$

$$
(a_{j_1 j_1 1}, x_{j_2 0}) = \underbrace{((2m_1)^{\text{th}}\text{LS,IC})}_{(2m_1+1)^{\text{th}}\text{-order lower-saddle}}. \tag{1.6}
$$

The flow at $(x_{j_1}^*, x_{j_2 0}) = (a_{j_1 j_1 1}, x_{j_2 0})$ *is a* $((2m_1)^{\text{th}}\text{LS,IC})$*-lower-saddle.*

(i_2) At $(x_{j_1}, x_{j_2}^*) = (\overline{x}_{j_1}, a_{j_1 j_2 1})$ *and* $\overline{x}_{j_1} \neq a_{j_1 j_1 1}$*, there are two cases* $n = 2n_1 + 1$ and $n = 2n_1$.

(i_{2a}) For $n = 2n_1 + 1$*, the equilibrium of* $(x_{j_1}, x_{j_2}^*) = (\overline{x}_{j_1}, a_{j_1 j_2 1})$ *and* $\overline{x}_{j_1} \neq a_{j_1 j_1 1}$ *has the following properties.*

- *For* $a_{j_1 j_1 0}(\overline{x}_{j_1} - a_{j_1 j_1 1})^m > 0$ *and* $a_{j_2 j_2 0} = 0$,

$$(\overline{x}_{j_1}, a_{j_1 j_2 1}) = \underbrace{(\text{PL}, (2\text{n}_1 + 1)^{\text{th}}\text{II})}_{(2n_1+1)^{\text{th}}\text{order increasing-inflection parallel flow}}. \tag{1.7}$$

The flow at $(x_{j_1}, x_{j_2}^*) = (\overline{x}_{j_1}, a_{j_1 j_2 1})$ *and* $\overline{x}_{j_1} \neq a_{j_1 j_1 1}$ *is a* $(\text{PL}, (2\text{n}_1 + 1)^{\text{th}}\text{II})$*-increasing-inflection parallel flow.*

- *For* $a_{j_1 j_1 0}(\overline{x}_{j_1} - a_{j_1 j_1 1})^m < 0$ *and* $a_{j_2 j_2 0} = 0$,

$$(\overline{x}_{j_1}, a_{j_1 j_2 1}) = \underbrace{(\text{PL}, (2\text{n}_1 + 1)^{\text{th}}\text{DI})}_{(2n_1+1)^{\text{th}}\text{order decreasing-inflection parallel flow}}. \tag{1.8}$$

The flow at $(x_{j_1}, x_{j_2}^*) = (\overline{x}_{j_1}, a_{j_1 j_2 1})$ *and* $\overline{x}_{j_1} \neq a_{j_1 j_1 1}$ *is a* $(\text{PL}, (2\text{n}_1 + 1)^{\text{th}}\text{DI})$*-increasing-inflection parallel flow.*

(i_{2b}) For $n = 2n_1$*, the equilibrium of* $(x_{j_1}, x_{j_2}^*) = (\overline{x}_{j_1}, a_{j_1 j_2 1})$ *and* $\overline{x}_{j_1} \neq a_{j_1 j_1 1}$ *has the following properties.*

- *For* $a_{j_1 j_1 0}(\overline{x}_{j_1} - a_{j_1 j_1 1})^m > 0$ *and* $a_{j_2 j_2 0} = 0$,

$$(\overline{x}_{j_1}, a_{j_1 j_2 1}) = \underbrace{(\text{PL}, (2\text{n}_1)^{\text{th}}\text{UP})}_{(2n_1)^{\text{th}}\text{order up-parabola parallel flow}}. \tag{1.9}$$

The flow at $(x_{j_1}, x_{j_2}^*) = (\overline{x}_{j_1}, a_{j_1 j_2 1})$ *and* $\overline{x}_{j_1} \neq a_{j_1 j_1 1}$ *is a* $(\text{PL}, (2\text{n}_1)^{\text{th}}\text{UP})$*-up-parabola parallel flow.*

- *For* $a_{j_1 j_1 0}(\overline{x}_{j_1} - a_{j_1 j_1 1})^m < 0$ *and* $a_{j_2 j_2 0} = 0$,

$$(\overline{x}_{j_1}, a_{j_1 j_2 1}) = \underbrace{(\text{PL}, (2\text{n}_1)^{\text{th}}\text{DP})}_{(2n_1)^{\text{th}}\text{order down-parabola parallel flow}}. \tag{1.10}$$

The flow at $(x_{j_1}, x_{j_2}^*) = (\overline{x}_{j_1}, a_{j_1 j_2 1})$ *and* $\overline{x}_{j_1} \neq a_{j_1 j_1 1}$ *is a* $(\text{PL}, (2\text{n}_1)^{\text{th}}\text{DP})$*-down-parabola parallel flow.*

(i_3) At $(x^*_{j_1}, x^*_{j_2}) = (a_{j_1 j_1 1}, a_{j_1 j_2 1})$, *there are four cases:* $m = 2m_1 + 1$ and $n = 2n_1 + 1$, $m = 2m_1 + 1$ and $n = 2n_1$; $m = 2m_1$ and $n = 2n_1 + 1$; and $m = 2m_1$ and $n = 2n_1$.

(i_{3a}) For $m = 2m_1 + 1$ and $n = 2n_1 + 1$, *the equilibrium of* $(x^*_{j_1}, x^*_{j_2}) = (a_{j_1 j_1 1}, a_{j_1 j_2 1})$ *has the following properties.*

- *For* $a_{j_1 j_1 0} > 0$ *and* $a_{j_2 j_2 0} = 0$,

$$(a_{j_1 j_1 1}, a_{j_1 j_2 1}) = \underbrace{((2m_1 + 1)^{\text{th}}\text{SI:SO},(2n_1 + 1)^{\text{th}}\text{DI:II})}_{[(2m_1+1),(2n_1+1)]\text{-inflection sink-to-source switch}}. \tag{1.11}$$

The flow at $(x^*_{j_1}, x^*_{j_2}) = (a_{j_1 j_1 1}, a_{j_1 j_2 1})$ *is a* $((2m_1 + 1)^{\text{th}}\text{SI:SO},(2n_1 + 1)^{\text{th}}\text{DI : II})$-*inflection sink-to-source switch.*

- *For* $a_{j_1 j_1 0} < 0$ *and* $a_{j_2 j_2 0} = 0$,

$$(a_{j_1 j_1 1}, a_{j_1 j_2 1}) = \underbrace{((2m_1 + 1)^{\text{th}}\text{SO:SI},(2n_1 + 1)^{\text{th}}\text{II:DI})}_{[(2m_1+1),(2n_1+1)]\text{-inflection source-to-sink switch}}. \tag{1.12}$$

The flow at $(x_{j_1}, x^*_{j_2}) = (\overline{x}_{j_1}, a_{j_1 j_2 1})$ *is a* $((2m_1 + 1)^{\text{th}}\text{SI:SO},(2n_1 + 1)^{\text{th}}\text{DI : II})$-*inflection source-to-sink switch.*

(i_{3b}) For $m = 2m_1 + 1$ and $n = 2n_1$, *the equilibrium of* $(x^*_{j_1}, x^*_{j_2}) = (a_{j_1 j_1 1}, a_{j_1 j_2 1})$ *has the following properties.*

- *For* $a_{j_1 j_1 0} > 0$ *and* $a_{j_2 j_2 0} = 0$,

$$(a_{j_1 j_1 1}, a_{j_1 j_2 1}) = \underbrace{((2m_1 + 1)^{\text{th}}\text{SO:SO},(2n_1)^{\text{th}}\text{DP:UP})}_{[(2m_1+1),(2n_1)]\text{-parabola sink-to-source switch}}. \tag{1.13}$$

The flow at $(x^*_{j_1}, x^*_{j_2}) = (a_{j_1 j_1 1}, a_{j_1 j_2 1})$ *is a* $((2m_1 + 1)^{\text{th}}\text{SO:SO},(2n_1 + 1)^{\text{th}}\text{DI : II})$-inflection source-to-source switch.

- *For* $a_{j_1 j_1 0} < 0$ *and* $a_{j_2 j_2 0} = 0$,

$$(a_{j_1 j_1 1}, a_{j_1 j_2 1}) = \underbrace{((2m_1 + 1)^{\text{th}}\text{SI:SI},(2n_1)^{\text{th}}\text{DP:UP})}_{[(2m_1+1),(2n_1)]\text{-parabola sink-to-sink switch}}. \tag{1.14}$$

The flow at $(x_{j_1}, x^*_{j_2}) = (\overline{x}_{j_1}, a_{j_1 j_2 1})$ *is a* $((2m_1 + 1)^{\text{th}}\text{SI:SI},(2n_1 + 1)^{\text{th}}\text{DI : II})$-*inflection sink-to-sink switch.*

(i_{3c}) For $m = 2m_1$ and $n = 2n_1 + 1$, *the equilibrium of* $(x^*_{j_1}, x^*_{j_2}) = (a_{j_1 j_1 1}, a_{j_1 j_2 1})$ *has the following properties.*

- *For* $a_{j_1j_10} > 0$ *and* $a_{j_2j_20} = 0$,

$$(a_{j_1j_11}, a_{j_1j_21}) = \underbrace{((2m_1)^{\text{th}}\text{LS:US},(2n_1+1)^{\text{th}}\text{II:II})}_{[(2m_1),(2n_1+1)]\text{-inflection lower-to-upper saddle switch}}. \tag{1.15}$$

The flow at $(x^*_{j_1}, x^*_{j_2}) = (a_{j_1j_11}, a_{j_1j_21})$ *is a* $((2m_1)^{\text{th}}\text{LS:US},(2n_1+1)^{\text{th}}\text{II : II})$-*inflection lower-to-upper saddle switch.*

- *For* $a_{j_1j_10} < 0$ *and* $a_{j_2j_20} = 0$,

$$(a_{j_1j_11}, a_{j_1j_21}) = \underbrace{((2m_1)^{\text{th}}\text{US:LS},(2n_1+1)^{\text{th}}\text{DI:DI})}_{[(2m_1),(2n_1+1)]\text{-inflection upper-to-lower saddle switch}}. \tag{1.16}$$

The flow at $(x_{j_1}, x^*_{j_2}) = (\overline{x}_{j_1}, a_{j_1j_21})$ *is a* $((2m_1)^{\text{th}}\text{US:LS},(2n_1+1)^{\text{th}}\text{DI : DI})$-*inflection upper-to-lower saddle switch.*

(i_{3d}) For $m = 2m_1$ and $n = 2n_1$, *the equilibrium of* $(x^*_{j_1}, x^*_{j_2}) = (a_{j_1j_11}, a_{j_1j_21})$ *has the following properties.*

- *For* $a_{j_1j_10} > 0$ *and* $a_{j_2j_20} = 0$,

$$(a_{j_1j_11}, a_{j_1j_21}) = \underbrace{((2m_1)^{\text{th}}\text{US:US},(2n_1)^{\text{th}}\text{UP:UP})}_{[(2m_1),(2n_1)]\text{-parabola upper-to-upper saddle switch}}. \tag{1.17}$$

The flow at $(x^*_{j_1}, x^*_{j_2}) = (a_{j_1j_11}, a_{j_1j_21})$ *is a* $((2m_1)^{\text{th}}\text{US:US},(2n_1)^{\text{th}}\text{UP : UP})$-*inflection upper-to-upper saddle switch.*

- *For* $a_{j_1j_10} < 0$ *and* $a_{j_2j_20} = 0$,

$$(a_{j_1j_11}, a_{j_1j_21}) = \underbrace{((2m_1)^{\text{th}}\text{US:US},(2n_1)^{\text{th}}\text{UP:UP})}_{[(2m_1),(2n_1)]\text{-parabola lower-to-lower saddle switch}}. \tag{1.18}$$

The flow at $(x_{j_1}, x^*_{j_2}) = (\overline{x}_{j_1}, a_{j_1j_21})$ *is a* $((2m_1)^{\text{th}}\text{LS:LS},(2n_1)^{\text{th}}\text{DP : DP})$ *inflection lower-to-lower saddle switch.*

(ii) If $a_{j_2j_20} \neq 0$, *the 2-dimensional dynamical system does not have any equilibriums for the* x_{j_2}*-direction on the* x_{j_1}*-direction. The standard form is*

$$\begin{aligned}\dot{x}_{j_2} &= a_{j_2j_20},\\ \dot{x}_{j_1} &= a_{j_1j_10}(x_{j_1} - a_{j_1j_11})^m (x_{j_2} - a_{j_1j_21})^n.\end{aligned} \tag{1.19}$$

The first integral manifold is

$$-\frac{1}{m-1}a_{j_2j_20}\left[\frac{1}{(x_{j_1}-a_{j_1j_11})^{m-1}}-\frac{1}{(x_{j_10}-a_{j_1j_11})^{m-1}}\right]$$
$$=\frac{1}{n+1}a_{j_1j_10}\left[(x_{j_2}-a_{j_1j_21})^{n+1}-(x_{j_20}-a_{j_1j_21})^{n+1}\right],$$
$$\text{for } m\neq 1;$$

$$a_{j_2j_20}\ln\frac{|x_{j_1}-a_{j_1j_11}|}{|x_{j_10}-a_{j_1j_11}|}=\tfrac{1}{n+1}a_{j_1j_10}\left[(x_{j_2}-a_{j_1j_21})^{n+1}-(x_{j_20}-a_{j_1j_21})^{n+1}\right],$$
$$\text{for } m=1. \tag{1.20}$$

There are four cases: (ii$_1$) $m=2m_1+1$ and $n=2n_1+1$, (ii$_2$) $m=2m_1$ and $n=2n_1+1$, (ii$_3$) $m=2m_1+1$ and $n=2n_1$, (ii$_4$) $m=2m_1$ and $n=2n_1$.

(ii$_1$) For $m=2m_1+1$ and $n=2n_1+1$, the equilibrium of $(x_{j_1}^*,x_{j_2}^*)=(a_{j_1j_11},a_{j_1j_21})$ for the x_{j_1}-direction has the following properties.

- *For $a_{j_1j_10}>0$ and $a_{j_2j_20}>0$,*

$$(a_{j_1j_11},a_{j_1j_21})=\underbrace{([(2m_1{+}1),(2n_1+1)]\text{DP:UP,pF})}_{[(2m_1+1),(2n_1+1)]\text{-hyperbolic flow }(+)}. \tag{1.21}$$

The flow at $(x_{j_1}^*,x_{j_2}^*)=(a_{j_1j_11},a_{j_1j_21})$ is a ([(2m_1+1), (2n_1 + 1)]DP:UP,pF)-*positive hyperbolic flow. The line of $x_{j_1}^*=a_{j_1j_11}$ is a* $(2m_1{+}1)^{\text{th}}$-order *positive sink-to-source directrix flow.*

- *For $a_{j_1j_10}<0$ and $a_{j_2j_20}>0$,*

$$(a_{j_1j_11},a_{j_1j_21})=\underbrace{([(2m_1{+}1),(2n_1+1)]\text{UP:DP,pF})}_{[(2m_1+1),(2n_1+1)]\text{-hyperbolic-secant flow }(+)}. \tag{1.22}$$

The flow at $(x_{j_1}^*,x_{j_2}^*)=(a_{j_1j_11},a_{j_1j_21})$ is a ([(2m_1+1), (2n_1 + 1)]UP:DP,pF)-positive *hyperbolic-secant flow. The line of $x_{j_1}^*=a_{j_1j_11}$ is a* $(2m_1{+}1)^{\text{th}}$-order *positive source-to-sink directrix flow.*

- *For $a_{j_1j_10}>0$ and $a_{j_2j_20}<0$,*

$$(a_{j_1j_11},a_{j_1j_21})=\underbrace{([(2m_1{+}1),(2n_1+1)]\text{UP:DP,nF})}_{[(2m_1+1),(2n_1+1)]\text{-hyperbolic-secant flow }(-)}. \tag{1.23}$$

The flow at $(x^*_{j_1}, x^*_{j_2}) = (a_{j_1j_11}, a_{j_1j_21})$ *is a* ([(2m_1+1), (2n_1 + 1)]UP:DP,nF)-negative *hyperbolic-secant flow. The line of* $x^*_{j_1} = a_{j_1j_11}$ *is a* $(2m_1+1)^{\text{th}}$-order *negative sink-to-source directrix flow.*

- *For* $a_{j_1j_10} < 0$ *and* $a_{j_2j_20} < 0$,

$$(a_{j_1j_11}, a_{j_1j_21}) = \underbrace{([(2m_1+1), (2n_1 + 1)]\text{UP:DP,nF})}_{[(2m_1+1),(2n_1+1)]\text{-hyperbolic-secant flow }(-)}. \quad (1.24)$$

The flow at $(x^*_{j_1}, x^*_{j_2}) = (a_{j_1j_11}, a_{j_1j_21})$ *is a* ([(2m_1+1), (2n_1 + 1)]UP:DP,nF)-*negative hyperbolic flow. The line of* $x^*_{j_1} = a_{j_1j_11}$ *is a* $(2m_1+1)^{\text{th}}$-order *negative source-to-sink directrix flow.*

(ii₂) For $m = 2m_1$ *and* $n = 2n_1 + 1$, *the equilibrium of* $(x^*_{j_1}, x^*_{j_2}) = (a_{j_1j_11}, a_{j_1j_21})$ *for the* x_{j_1}*-direction has the following properties.*

- *For* $a_{j_1j_10} > 0$ *and* $a_{j_2j_20} > 0$,

$$(a_{j_1j_11}, a_{j_1j_21}) = \underbrace{([(2m_1), (2n_1 + 1)]\text{UP:UP,pF})}_{[(2m_1),(2n_1+1)]\text{ - hyperbolic-secant-to-hyperbolic flow }(+)}. \quad (1.25)$$

The flow at $(x^*_{j_1}, x^*_{j_2}) = (a_{j_1j_11}, a_{j_1j_21})$ *is a* ([(2m_1), (2n_1 + 1)]UP:UP,pF)-*positive hyperbolic-secant to hyperbolic flow. The line of* $x^*_{j_1} = a_{j_1j_11}$ *is a* $(2m_1)^{\text{th}}$*-order positive lower-to-upper saddle directrix flow.*

- *For* $a_{j_1j_10} < 0$ *and* $a_{j_2j_20} > 0$,

$$(a_{j_1j_11}, a_{j_1j_21}) = \underbrace{([(2m_1), (2n_1 + 1)]\text{DP:DP,pF})}_{[(2m_1),(2n_1+1)]\text{-hyperbolic-to-hyperbolic-secant flow }(+)}. \quad (1.26)$$

The flow at $(x^*_{j_1}, x^*_{j_2}) = (a_{j_1j_11}, a_{j_1j_21})$ *is a* ([(2m_1), (2n_1 + 1)]DP:DP,pF)-positive hyperbolic to *hyperbolic-secant flow. The line of* $x^*_{j_1} = a_{j_1j_11}$ *is a* $(2m_1)^{\text{th}}$*-order positive upper-to-lower saddle directrix flow.*

- *For* $a_{j_1j_10} > 0$ *and* $a_{j_2j_20} < 0$,

$$(a_{j_1j_11}, a_{j_1j_21}) = \underbrace{([(2m_1), (2n_1 + 1)]\text{DP:DP,nF})}_{[(2m_1),(2n_1+1)]\text{-hyperbolic-to-hyperbolic-secant flow }(-)}. \quad (1.27)$$

The flow at $(x^*_{j_1}, x^*_{j_2}) = (a_{j_1j_11}, a_{j_1j_21})$ *is a* ([(2m_1), (2n_1 + 1)]DP:DP,nF)-negative *hyperbolic-to-hyperbolic-secant flow. The line of* $x^*_{j_1} = a_{j_1j_11}$ *is a* $(2m_1)^{\text{th}}$-order *negative lower-to-upper saddle directrix flow.*

- *For* $a_{j_1j_10} < 0$ *and* $a_{j_2j_20} < 0$,

$$(a_{j_1j_11}, a_{j_1j_21}) = \underbrace{([(2m_1), (2n_1+1)]\text{UP:UP,nF})}_{[(2m_1),(2n_1+1)]\text{-hyperbolic-secant-to-hyperbolic flow }(-)}. \quad (1.28)$$

The flow at $(x^*_{j_1}, x^*_{j_2}) = (a_{j_1j_11}, a_{j_1j_21})$ *is a* $([(2m_1), (2n_1+1)]\text{UP:UP,nF})$*-negative hyperbolic-secant-to-hyperbolic flow. The line of* $x^*_{j_1} = a_{j_1j_11}$ *is a* $(2m_1)^{\text{th}}$*-order negative upper-to-lower saddle directrix flow.*

(ii₃) For $m = 2m_1 + 1$ *and* $n = 2n_1$*, the equilibrium of* $(x^*_{j_1}, x^*_{j_2}) = (a_{j_1j_11}, a_{j_1j_21})$ *for the* x_{j_1}*-direction has the following properties.*

- *For* $a_{j_1j_10} > 0$ *and* $a_{j_2j_20} > 0$,

$$(a_{j_1j_11}, a_{j_1j_21}) = \underbrace{([(2m_1+1), (2n_1)]_{\text{DI:II}}\text{SO,pF})}_{[(2m_1+1),(2n_1)]\text{-decreasing-to-increasing inflection source flow }(+)}. \quad (1.29)$$

The flow at $(x^*_{j_1}, x^*_{j_2}) = (a_{j_1j_11}, a_{j_1j_21})$ *is a* $([(2m_1+1), (2n_1)]_{\text{DI:II}}\text{SO,pF})$*-positive decreasing-to-increasing inflection source flow. The line of* $x^*_{j_1} = a_{j_1j_11}$ *is a* $(2m_1+1)^{\text{th}}$*-order positive source directrix flow.*

- *For* $a_{j_1j_10} < 0$ *and* $a_{j_2j_20} > 0$,

$$(a_{j_1j_11}, a_{j_1j_21}) = \underbrace{([(2m_1+1), (2n_1)]_{\text{II:DI}}\text{SI,pF})}_{[(2m_1+1),(2n_1)]\text{-increasing-to-decreasing inflection sink flow }(+)}. \quad (1.30)$$

The flow at $(x^*_{j_1}, x^*_{j_2}) = (a_{j_1j_11}, a_{j_1j_21})$ *is a* $([(2m_1+1), (2n_1)]_{\text{II:DI}}\text{SI,pF})$*-positive increasing-to-decreasing inflection source flow. The line of* $x^*_{j_1} = a_{j_1j_11}$ *is a* $(2m_1+1)^{\text{th}}$*-order positive sink directrix flow.*

- *For* $a_{j_1j_10} > 0$ *and* $a_{j_2j_20} < 0$,

$$(a_{j_1j_11}, a_{j_1j_21}) = \underbrace{([(2m_1+1), (2n_1)]_{\text{II:DI}}\text{SO,nF})}_{[(2m_1+1),(2n_1)]\text{-increasing-to-decreasing inflection source flow }(-)}. \quad (1.31)$$

The flow at $(x^*_{j_1}, x^*_{j_2}) = (a_{j_1j_11}, a_{j_1j_21})$ *is a* $([(2m_1+1), (2n_1)]_{\text{II:DI}}\text{SO,nF})$*-negative increasing-to-decreasing inflection source flow. The line of* $x^*_{j_1} = a_{j_1j_11}$ *is a* $(2m_1+1)^{\text{th}}$*-order negative source directrix flow.*

- *For* $a_{j_1j_10} < 0$ *and* $a_{j_2j_20} < 0$,

$$(a_{j_1j_11}, a_{j_1j_21}) = \underbrace{([(2m_1+1), (2n_1)]_{\text{DI:II}}\text{SI,nF})}_{[(2m_1+1),(2n_1)]\text{-decreasing-to-increasing inflection sink flow }(-)}. \quad (1.32)$$

The flow at $(x^*_{j_1}, x^*_{j_2}) = (a_{j_1j_11}, a_{j_1j_21})$ *is a* ([(2m_1+1), (2n_1)]$_{\text{DI:II}}$SI,nF)-*negative decreasing-to-increasing inflection sink flow. The line of* $x^*_{j_1} = a_{j_1j_11}$ *is a* $(2m_1+1)^{\text{th}}$-order *negative sink directrix flow.*

(ii₄) For $m = 2m_1$ *and* $n = 2n_1$*, the equilibrium of* $(x^*_{j_1}, x^*_{j_2}) = (a_{j_1j_11}, a_{j_1j_21})$ *for the* x_{j_1}*-direction has the following properties.*

- *For* $a_{j_1j_10} > 0$ *and* $a_{j_2j_20} > 0$,

$$(a_{j_1j_11}, a_{j_1j_21}) = \underbrace{([(2m_1), (2n_1)]_{\text{II:II}}\text{US,pF})}_{[(2m_1),(2n_1)]\text{-increasing-inflection upper-saddle flow }(+)}. \tag{1.33}$$

The flow at $(x^*_{j_1}, x^*_{j_2}) = (a_{j_1j_11}, a_{j_1j_21})$ *is a* ([(2m_1), (2n_1)]$_{\text{II:II}}$US,pF)-*positive increasing-inflection upper-saddle flow. The line of* $x^*_{j_1} = a_{j_1j_11}$ *is a* $(2m_1)^{\text{th}}$-*order positive upper-saddle directrix flow.*

- *For* $a_{j_1j_10} < 0$ *and* $a_{j_2j_20} > 0$,

$$(a_{j_1j_11}, a_{j_1j_21}) = \underbrace{([(2m_1), (2n_1)]_{\text{DI:DI}}\text{LS,pF})}_{[(2m_1),(2n_1)]\text{-decreasing-inflection lower-saddle flow }(+)}. \tag{1.34}$$

The flow at $(x^*_{j_1}, x^*_{j_2}) = (a_{j_1j_11}, a_{j_1j_21})$ *is a* ([(2m_1), (2n_1)]$_{\text{DI:DI}}$LS,pF)-*positive decreasing-inflection lower-saddle flow. The line of* $x^*_{j_1} = a_{j_1j_11}$ *is a* $(2m_1)^{\text{th}}$-order *positive lower-saddle directrix flow.*

- *For* $a_{j_1j_10} > 0$ *and* $a_{j_2j_20} < 0$,

$$(a_{j_1j_11}, a_{j_1j_21}) = \underbrace{([(2m_1), (2n_1)]_{\text{DI:DI}}\text{US,nF})}_{[(2m_1),(2n_1)]\text{-decreasing-inflection upper-saddle flow }(-)}. \tag{1.35}$$

The flow at $(x^*_{j_1}, x^*_{j_2}) = (a_{j_1j_11}, a_{j_1j_21})$ *is a* ([(2m_1), (2n_1)]$_{\text{DI:DI}}$US,nF)-*negative decreasing-inflection upper-saddle flow*([(2m_1), (2n_1)]$_{\text{DI:DI}}$US,nF). *The line of* $x^*_{j_1} = a_{j_1j_11}$ *is a* $(2m_1)^{\text{th}}$-order *negative upper-saddle directrix flow.*

- *For* $a_{j_1j_10} < 0$ *and* $a_{j_2j_20} < 0$,

$$(a_{j_1j_11}, a_{j_1j_21}) = \underbrace{([(2m_1), (2n_1)]_{\text{II:II}}\text{LS,nF})}_{[(2m_1),(2n_1)]\text{-increasing-inflection lower-saddle flow }(-)}. \tag{1.36}$$

The flow at $(x^*_{j_1}, x^*_{j_2}) = (a_{j_1j_11}, a_{j_1j_21})$ *is a* ([(2m_1), (2n_1)]$_{\text{II:II}}$LS,nF)-*negative increasing-inflection lower-saddle flow. The line of* $x^*_{j_1} = a_{j_1j_11}$ *is a* $(2m_1)^{\text{th}}$-order *positive lower-saddle directrix flow.*

(iii) Consider

$$\begin{aligned}\dot{x}_{j_2} &= a_{j_2j_20}\\ \dot{x}_{j_1} &= a_{j_1j_10}\prod\nolimits_{\alpha=1}^{p}(x_{j_1}-a_{j_1j_1\alpha})^{m_\alpha}\prod\nolimits_{\beta=1}^{q}(x_{j_2}-a_{j_1j_2\beta})^{m_\beta},\end{aligned} \tag{1.37}$$

where

$$\sum\nolimits_{\alpha=1}^{p} m_\alpha = m, \sum\nolimits_{\beta=1}^{q} n_\beta = n. \tag{1.38}$$

The first integral manifold is

$$\begin{aligned}
&a_{j_2j_20}\sum_{\alpha_1=1}^{p}\sum_{l_1=1}^{m_{\alpha_1}}\frac{A_{l_1m_{\alpha_1}}}{l_1-1}\Big[\frac{1}{(x_{j_1}-a_{j_1j_1\alpha_1})^{l_1-1}}-\frac{1}{(x_{j_10}-a_{j_1j_1\alpha_1})^{l_1-1}}\Big]\\
&= a_{j_1j_10}\Big\{\frac{1}{n+1}\big[(x_{j_2}-a_{j_1j_2\beta_1})^{n+1}-(x_{j_20}-a_{j_1j_2\beta_1})^{n+1}\big]\\
&+\sum_{\rho=1}^{n-n_{\beta_1}}\frac{b_{j_1j_2\rho}}{n-\rho+1}\big[(x_{j_2}-a_{j_1j_2\beta_1})^{n-\rho+1}-(x_{j_20}-a_{j_1j_2\beta_1})^{n-\rho+1}\big]\Big\},\\
&\text{for } l_1\neq 1;\\
&\lim_{l_1\to 1}\frac{1}{l_1-1}\Big[\frac{1}{(x_{j_1}-a_{j_1j_1\alpha_1})^{l_1-1}}-\frac{1}{(x_{j_10}-a_{j_1j_1\alpha_1})^{l_1-1}}\Big]=-\ln\frac{|x_{j_1}-a_{j_1j_1\alpha_1}|}{|x_{j_10}-a_{j_1j_1\alpha_1}|},\\
&\text{for } l_1=1;
\end{aligned} \tag{1.39}$$

where

$$\begin{aligned}
A_{l_1m_{\alpha_1}} &= \frac{1}{(m_{r_1}-l_1)!}\frac{d^{m_{\alpha_1}-l_1}}{dx_{j_1}^{m_{\alpha_1}-l_1}}\frac{1}{\prod\nolimits_{\alpha_2=1,\alpha_2\neq\alpha_1}^{p}(x_{j_1}-a_{j_1j_1\alpha_2})^{m_{\alpha_2}}}\Big|_{x_{j_1}^*=a_{j_1j_1\alpha_1}},\\
A_{m_{\alpha_1}m_{\alpha_1}} &= \frac{1}{\prod\nolimits_{\alpha_2=1,\alpha_2\neq\alpha_1}^{p}(a_{j_1j_1\alpha_1}-a_{j_1j_1\alpha_2})^{m_{\alpha_2}}};\\
b_{j_1j_2\rho} &= \sum_{l_{\beta_q},t_{\beta_q}=0}^{l_{\beta_q}+t_{\beta_q}=n_{\beta_q}}\cdots\sum_{l_{\beta_2},t_{\beta_2}=0}^{l_{\beta_2}+t_{\beta_2}=n_{\beta_2}}\Big[\prod_{k=\beta_2}^{\beta_q}\frac{n_k!}{l_k!t_k!}(a_{j_1j_2\beta_1}-a_{j_1j_2k})^{l_k}\Big]\delta_{t_{\beta_2}+\cdots+t_{\beta_q}}^{n-\rho},\\
b_{j_1j_20} &= 1.
\end{aligned} \tag{1.40}$$

There are four cases: *(iii$_1$)* $m_{\alpha_1}=2m_{\alpha_1 1}+1$ *and* $n_{\beta_1}=2n_{\beta_1 1}+1$, *(iii$_2$)* $m_{\alpha_1}=2m_{\alpha_1 1}$ *and* $n_{\beta_1}=2n_{\beta_1 1}+1$, *(iii$_3$)* $m_{\alpha_1}=2m_{\alpha_1 1}+1$ *and* $n_{\beta_1}=2n_{\beta_1 1}$, *(iii$_4$)* $m_{\alpha_1}=2m_{\alpha_1 1}$ *and* $n_{\beta_1}=2n_{\beta_1 1}$. *Define*

$$\bigcup_{\alpha=1}^{p}\bigcup_{\beta=1}^{q}\underbrace{(a_{j_1j_1\alpha},a_{j_1j_2\beta})}_{([m_\alpha,m_\beta]\text{XX,xF})}$$

$$=\begin{bmatrix}\underbrace{(a_{j_1j_11},a_{j_1j_2q})}_{([m_1,n_q]XX,xF)} & \cdots & \underbrace{(a_{j_1j_1(p-1)},a_{j_1j_2q})}_{([m_{p-1},n_q]XX,xF)} & \underbrace{(a_{j_1j_1p},a_{j_1j_2q})}_{([m_p,n_q]XX,xF)}\\ \underbrace{(a_{j_1j_11},a_{j_1j_2(q-1)})}_{([m_1,n_q]XX,xF)} & \cdots & \underbrace{(a_{j_1j_1(p-1)},a_{j_1j_2(q-1)})}_{([m_{p-1},n_{q-1}]XX,xF)} & \underbrace{(a_{j_1j_1p},a_{j_1j_2(q-1)})}_{([m_p,n_{q-1}]XX,xF)}\\ \vdots & \vdots & \vdots & \vdots\\ \underbrace{(a_{j_1j_11},a_{j_1j_21})}_{([m_1,n_1]XX,xF)} & \cdots & \underbrace{(a_{j_1j_1(p-1)},a_{j_1j_21})}_{([m_{p-1},n_1]XX,xF)} & \underbrace{(a_{j_1j_1p_{j_1}},a_{j_1j_21})}_{([m_{p_{j_1}},n_1]XX,xF)}\end{bmatrix}. \tag{1.41}$$

(iii_1) For $m_{\alpha_1}=2m_{\alpha_11}+1$ and $n_{\beta_1}=2n_{\beta_11}+1$, the equilibriums of $(x_{j_1}^*,x_{j_2}^*)=(a_{j_1j_1\alpha_1},a_{j_1j_2\beta_1})(\alpha_1\in\{1,2,\cdots,p\};\ \beta_1\in\{1,2,\cdots,q\})$ for the x_{j_1}-direction has the following properties.

- *For $a_{j_1j_10}\prod_{\alpha_1=1,\alpha_2\neq\alpha_1}^{p}(a_{j_1j_1\alpha_1}-a_{j_1j_1\alpha_2})^{m_{\alpha_2}}\prod_{\beta_2=1,\beta_2\neq\beta_1}^{q}(a_{j_1j_2\beta_1}-a_{j_1j_2\beta_2})^{n_{\beta_2}}>0$ and $a_{j_2j_20}>0$,*

$$(a_{j_1j_1\alpha_1},a_{j_1j_2\beta_1})=\underbrace{([(2m_{\alpha_11}+1),\,(2n_{\beta_11}+1)]\text{DP:UP,pF})}_{[(2m_{\alpha_11}+1),(2n_{\beta_11}+1)]\text{-hyperbolic flow }(+)}. \tag{1.42}$$

The flow at $(x_{j_1}^*,x_{j_2}^*)=(a_{j_1j_1\alpha_1},a_{j_1j_2\beta_1})$ is a ([$(2m_{\alpha_11}+1)$, $(2n_{\beta_11}+1)$]DP:UP,pF)-positive hyperbolic flow. The line of $x_{j_1}^*=a_{j_1j_1\alpha_1}$ is a $(2m_{\alpha_11}+1)^{\text{th}}$-order positive sink-to-source directrix flow.

- *For $a_{j_1j_10}\prod_{\alpha_1=1,\alpha_2\neq\alpha_1}^{p}(a_{j_1j_1\alpha_1}-a_{j_1j_1\alpha_2})^{m_{\alpha_2}}\prod_{\beta_2=1,\beta_2\neq\beta_1}^{q}(a_{j_1j_2\beta_1}-a_{j_1j_2\beta_2})^{n_{\beta_2}}<0$ and $a_{j_2j_20}>0$,*

$$(a_{j_1j_1\alpha_1},a_{j_1j_2\beta_1})=\underbrace{([(2m_{\alpha_11}+1),\,(2n_{\beta_11}+1)]\text{UP:DP,pF})}_{[(2m_{\alpha_11}+1),(2n_{\beta_11}+1)]\text{-hyperbolic-secant flow }(+)}. \tag{1.43}$$

The flow at $(x_{j_1}^*,x_{j_2}^*)=(a_{j_1j_1\alpha_1},a_{j_1j_2\beta_1})$ is a ([$(2m_{\alpha_11}+1)$, $(2n_{\beta_11}+1)$]UP:DP,pF)-positive hyperbolic-secant flow. The line of $x_{j_1}^*=a_{j_1j_1\alpha_1}$ is a $(2m_{\alpha_11}+1)^{\text{th}}$-order positive source-to-sink directrix flow.

- *For $a_{j_1j_10}\prod_{\alpha_1=1,\alpha_2\neq\alpha_1}^{p}(a_{j_1j_1\alpha_1}-a_{j_1j_1\alpha_2})^{m_{\alpha_2}}\prod_{\beta_2=1,\beta_2\neq\beta_1}^{q}(a_{j_1j_2\beta_1}-a_{j_1j_2\beta_2})^{n_{\beta_2}}>0$ and $a_{j_2j_20}<0$,*

$$(a_{j_1j_1\alpha_1},a_{j_1j_2\beta_1})=\underbrace{([(2m_{\alpha_11}+1),\,(2n_{\beta_11}+1)]\text{UP:DP,nF})}_{[(2m_{\alpha_11}+1),(2n_{\beta_11}+1)]\text{-hyperbolic-secant flow }(-)}. \tag{1.44}$$

The flow at $(x_{j_1}^*, x_{j_2}^*) = (a_{j_1j_1\alpha_1}, a_{j_1j_2\beta_1})$ *is a* ([$(2m_{\alpha_1 1}+1), (2n_{\beta_1 1}+1)$]UP:DP,nF) *-negative hyperbolic-secant flow. The line of* $x_{j_1}^* = a_{j_1j_1\alpha_1}$ *is a* $(2m_{\alpha_1 1}+1)^{\text{th}}$-order *negative sink-to-source directrix flow.*

- *For* $a_{j_1j_10}\prod_{\alpha_1=1,\alpha_2\neq\alpha_1}^{p}(a_{j_1j_1\alpha_1}-a_{j_1j_1\alpha_2})^{m_{\alpha_2}}\prod_{\beta_2=1,\beta_2\neq\beta_1}^{q}(a_{j_1j_2\beta_1}-a_{j_1j_2\beta_2})^{n_{\beta_2}} < 0$ *and* $a_{j_2j_20} < 0$,

$$(a_{j_1j_1\alpha_1}, a_{j_1j_2\beta_1}) = \underbrace{([(2m_{\alpha_1 1}+1), (2n_{\beta_1 1}+1)]\text{DP:UP,nF})}_{[(2m_{\alpha_1 1}+1),(2n_{\beta_1 1}+1)]\text{-hyperbolic flow }(-)}. \quad (1.45)$$

The flow at $(x_{j_1}^*, x_{j_2}^*) = (a_{j_1j_1\alpha_1}, a_{j_1j_2\beta_1})$ *is a* ([$(2m_{\alpha_1 1}+1), (2n_{\beta_1 1}+1)$]DP:UP,nF)*-negative hyperbolic flow. The line of* $x_{j_1}^* = a_{j_1j_1\alpha_1}$ *is a* $(2m_{\alpha_1 1}+1)^{\text{th}}$*-order negative source-to-sink directrix flow.*

(iii₂) For $m_{\alpha_1} = 2m_{\alpha_1 1}$ *and* $n_{\beta_1} = 2n_{\beta_1 1}+1$, *the equilibriums of* $(x_{j_1}^*, x_{j_2}^*) = (a_{j_1j_1\alpha_1}, a_{j_1j_2\beta_1})$ $(\alpha_1 \in \{1,2,\cdots,p\};\ \beta_1 \in \{1,2,\cdots,q\})$ *for the* x_{j_1}*-direction has the following properties.*

- *For* $a_{j_1j_10}\prod_{\alpha_1=1,\alpha_2\neq\alpha_1}^{p}(a_{j_1j_1\alpha_1}-a_{j_1j_1\alpha_2})^{m_{\alpha_2}}\prod_{\beta_2=1,\beta_2\neq\beta_1}^{q}(a_{j_1j_2\beta_1}-a_{j_1j_2\beta_2})^{n_{\beta_2}} > 0$ *and* $a_{j_2j_20} > 0$,

$$(a_{j_1j_1\alpha_1}, a_{j_1j_2\beta_1}) = \underbrace{([(2m_{\alpha_1 1}), (2n_{\beta_1 1}+1)]\text{UP:UP,pF})}_{[(2m_{\alpha_1 1}),(2n_{\beta_1 1}+1)]\text{-hyperbolic-secant-to-hyperbolic flow }(+)}. \quad (1.46)$$

The flow at $(x_{j_1}^*, x_{j_2}^*) = (a_{j_1j_1\alpha_1}, a_{j_1j_2\beta_1})$ *is a* ([$(2m_{\alpha_1 1}), (2n_{\beta_1 1}+1)$]UP:UP,pF)*-positive hyperbolic-secant-to-hyperbolic flow. The line of* $x_{j_1}^* = a_{j_1j_1\alpha_1}$ *is a* $(2m_{\alpha_1 1})^{\text{th}}$*-order positive lower-to-upper saddle directrix flow.*

- *For* $a_{j_1j_10}\prod_{\alpha_1=1,\alpha_2\neq\alpha_1}^{p}(a_{j_1j_1\alpha_1}-a_{j_1j_1\alpha_2})^{m_{\alpha_2}}\prod_{\beta_2=1,\beta_2\neq\beta_1}^{q}(a_{j_1j_2\beta_1}-a_{j_1j_2\beta_2})^{n_{\beta_2}} < 0$ *and* $a_{j_2j_20} > 0$,

$$(a_{j_1j_1\alpha_1}, a_{j_1j_2\beta_1}) = \underbrace{([(2m_{\alpha_1 1}), (2n_{\beta_1 1}+1)]\text{DP:DP,pF})}_{[(2m_{\alpha_1 1}),(2n_{\beta_1 1}+1)]\text{-hyperbolic-to-hyperbolic-secant flow }(+)}. \quad (1.47)$$

The flow at $(x_{j_1}^*, x_{j_2}^*) = (a_{j_1j_1\alpha_1}, a_{j_1j_2\beta_1})$ *is a* ([$(2m_{\alpha_1 1}), (2n_{\beta_1 1}+1)$]UP:UP,pF)*-positive hyperbolic-to-hyperbolic-secant flow. The line of* $x_{j_1}^* = a_{j_1j_1\alpha_1}$ *is a* $(2m_{\alpha_1 1})^{\text{th}}$*-order positive upper-to-lower saddle directrix flow.*

- *For* $a_{j_1j_10}\prod_{\alpha_1=1,\alpha_2\neq\alpha_1}^{p}(a_{j_1j_1\alpha_1}-a_{j_1j_1\alpha_2})^{m_{\alpha_2}}\prod_{\beta_2=1,\beta_2\neq\beta_1}^{q}(a_{j_1j_2\beta_1}-a_{j_1j_2\beta_2})^{n_{\beta_2}} > 0$ *and* $a_{j_2j_20} < 0$,

$$(a_{j_1j_1\alpha_1}, a_{j_1j_2\beta_1}) = \underbrace{([(2m_{\alpha_1 1}), (2n_{\beta_1 1}+1)]\text{DP:DP,nF})}_{[(2m_{\alpha_1 1}),(2n_{\beta_1 1}+1)]\text{-hyperbolic-to-hyperbolic-secant flow }(-)}. \quad (1.48)$$

The flow at $(x_{j_1}^*, x_{j_2}^*) = (a_{j_1 j_1 \alpha_1}, a_{j_1 j_2 \beta_1})$ *is a* ([$(2m_{\alpha_1 1}), (2n_{\beta_1 1}+1)$]DP:DP,nF)-*negative hyperbolic-to-hyperbolic-secant flow. The line of* $x_{j_1}^* = a_{j_1 j_1 \alpha_1}$ *is a* $(2m_{\alpha_1 1})^{\text{th}}$-order *negative lower-to-upper saddle directrix flow.*

- *For* $a_{j_1 j_1 0} \prod_{\alpha_1=1,\alpha_2\neq\alpha_1}^{p} (a_{j_1 j_1 \alpha_1} - a_{j_1 j_1 \alpha_2})^{m_{\alpha_2}} \prod_{\beta_2=1,\beta_2\neq\beta_1}^{q} (a_{j_1 j_2 \beta_1} - a_{j_1 j_2 \beta_2})^{n_{\beta_2}} < 0$ *and* $a_{j_2 j_2 0} < 0$,

$$(a_{j_1 j_1 \alpha_1}, a_{j_1 j_2 \beta_1}) = \underbrace{([(2m_{\alpha_1 1}), (2n_{\beta_1 1}+1)]\text{UP:UP,nF})}_{[(2m_{\alpha_1 1}),(2n_{\beta_1 1}+1)]\text{-hyperbolic-secant-to-hyperbolic flow }(-)}. \tag{1.49}$$

The flow at $(x_{j_1}^*, x_{j_2}^*) = (a_{j_1 j_1 \alpha_1}, a_{j_1 j_2 \beta_1})$ *is a* ([$(2m_{\alpha_1 1}), (2n_{\beta_1 1}+1)$]UP:UP,nF)-*negative hyperbolic-secant-to-hyperbolic flow. The line of* $x_{j_1}^* = a_{j_1 j_1 \alpha_1}$ *is a* $(2m_{\alpha_1 1})^{\text{th}}$-order *negative upper-to-lower saddle directrix flow.*

(iii$_3$*) For* $m_{\alpha_1} = 2m_{\alpha_1 1} + 1$ *and* $n_{\beta_1} = 2n_{\beta_1 1}$, *the equilibriums of* $(x_{j_1}^*, x_{j_2}^*) = (a_{j_1 j_1 \alpha_1}, a_{j_1 j_2 \beta_1})(\alpha_1 \in \{1, 2, \cdots, p\}; \beta_1 \in \{1, 2, \cdots, q\})$ *for the* x_{j_1}*-direction has the following properties.*

- *For* $a_{j_1 j_1 0} \prod_{\alpha_1=1,\alpha_2\neq\alpha_1}^{p} (a_{j_1 j_1 \alpha_1} - a_{j_1 j_1 \alpha_2})^{m_{\alpha_2}} \prod_{\beta_2=1,\beta_2\neq\beta_1}^{q} (a_{j_1 j_2 \beta_1} - a_{j_1 j_2 \beta_2})^{n_{\beta_2}} > 0$ *and* $a_{j_2 j_2 0} > 0$,

$$(a_{j_1 j_1 \alpha_1}, a_{j_1 j_2 \beta_1}) = \underbrace{([(2m_{\alpha_1 1}+1), (2n_{\beta_1 1})]_{\text{DI:II}}\text{SO,pF})}_{[(2m_{\alpha_1 1}+1),(2n_{\beta_1 1})]\text{-increasing-to-decreasing inflection source flow }(+)}. \tag{1.50}$$

The flow at $(x_{j_1}^*, x_{j_2}^*) = (a_{j_1 j_1 \alpha_1}, a_{j_1 j_2 \beta_1})$ *is a* ([$(2m_{\alpha_1 1}+1), (2n_{\beta_1 1})$]$_{\text{DI:II}}$SO,pF)-*positive decreasing-to-increasing inflection source flow. The line of* $x_{j_1}^* = a_{j_1 j_1 \alpha_1}$ $x_{j_1}^* = a_{j_1 j_1 \alpha_1}$ *is a* $(2m_{\alpha_1 1}+1)^{\text{th}}$-order positive *source directrix flow.*

- *For* $a_{j_1 j_1 0} \prod_{\alpha_1=1,\alpha_2\neq\alpha_1}^{p} (a_{j_1 j_1 \alpha_1} - a_{j_1 j_1 \alpha_2})^{m_{\alpha_2}} \prod_{\beta_2=1,\beta_2\neq\beta_1}^{q} (a_{j_1 j_2 \beta_1} - a_{j_1 j_2 \beta_2})^{n_{\beta_2}} < 0$ *and* $a_{j_2 j_2 0} > 0$,

$$(a_{j_1 j_1 \alpha_1}, a_{j_1 j_2 \beta_1}) = \underbrace{([(2m_{\alpha_1 1}+1), (2n_{\beta_1 1})]_{\text{II:DI}}\text{SI,pF})}_{[(2m_{\alpha_1 1}+1),(2n_{\beta_1 1})]\text{-decreasing-to-increasing inflection sink flow }(+)}. \tag{1.51}$$

The flow at $(x_{j_1}^*, x_{j_2}^*) = (a_{j_1 j_1 \alpha_1}, a_{j_1 j_2 \beta_1})$ *is a* ([$(2m_{\alpha_1 1}+1), (2n_{\beta_1 1})$]$_{\text{II:DI}}$SI,pF)-*positive increasing-to-decreasing inflection sink flow. The line of* $x_{j_1}^* = a_{j_1 j_1 \alpha_1}$ *is a* $(2m_{\alpha_1 1}+1)^{\text{th}}$-order positive *sink directrix flow.*

- *For* $a_{j_1 j_1 0} \prod_{\alpha_1=1,\alpha_2\neq\alpha_1}^{p} (a_{j_1 j_1 \alpha_1} - a_{j_1 j_1 \alpha_2})^{m_{\alpha_2}} \prod_{\beta_2=1,\beta_2\neq\beta_1}^{q} (a_{j_1 j_2 \beta_1} - a_{j_1 j_2 \beta_2})^{n_{\beta_2}} > 0$ *and* $a_{j_2 j_2 0} < 0$,

$$(a_{j_1 j_1 \alpha_1}, a_{j_1 j_2 \beta_1}) = \underbrace{([(2m_{\alpha_1 1}+1), (2n_{\beta_1 1})]_{\text{II:DI}}\text{SO,nF})}_{[(2m_{\alpha_1 1}+1),(2n_{\beta_1 1})]\text{-decreasing-to-increasing inflection source flow }(-)}. \tag{1.52}$$

The flow at $(x^*_{j_1}, x^*_{j_2}) = (a_{j_1 j_1 \alpha_1}, a_{j_1 j_2 \beta_1})$ *is a* $([(2m_{\alpha_1 1}+1), (2n_{\beta_1 1})]_{\text{II:DI}}\text{SO,nF})$-*negative increasing-to-decreasing inflection source flow. The line of* $x^*_{j_1} = a_{j_1 j_1 \alpha_1}$ *is a* $(2m_{\alpha_1 1}+1)^{\text{th}}$-order negative *source directrix flow.*

- *For* $a_{j_1 j_1 0} \prod_{\alpha_1=1,\alpha_2 \neq \alpha_1}^{p} (a_{j_1 j_1 \alpha_1} - a_{j_1 j_1 \alpha_2})^{m_{\alpha_2}} \prod_{\beta_2=1,\beta_2 \neq \beta_1}^{q} (a_{j_1 j_2 \beta_1} - a_{j_1 j_2 \beta_2})^{n_{\beta_2}} < 0$ *and* $a_{j_2 j_2 0} < 0$,

$$(a_{j_1 j_1 \alpha_1}, a_{j_1 j_2 \beta_1}) = \underbrace{([(2m_{\alpha_1 1}+1), (2n_{\beta_1 1})]_{\text{DI:II}}\text{SI,nF})}_{[(2m_{\alpha_1 1}+1),(2n_{\beta_1 1})]\text{-increasing-to-decreasing inflection sink flow }(-)}. \quad (1.53)$$

The flow at $(x^*_{j_1}, x^*_{j_2}) = (a_{j_1 j_1 \alpha_1}, a_{j_1 j_2 \beta_1})$ *is a* $([(2m_{\alpha_1 1}+1), (2n_{\beta_1 1})]_{\text{DI:II}}\text{SI,nF})$-*negative decreasing-to-increasing inflection sink flow. The line of* $x^*_{j_1} = a_{j_1 j_1 \alpha_1}$ *is a* $(2m_{\alpha_1 1}+1)^{\text{th}}$-order negative *sink directrix flow.*

(iii₄) For $m_{\alpha_1} = 2m_{\alpha_1 1}$ *and* $n_{\beta_1} = 2n_{\beta_1 1}$, *the equilibriums of* $(x^*_{j_1}, x^*_{j_2}) = (a_{j_1 j_1 \alpha_1}, a_{j_1 j_2 \beta_1})(\alpha_1 \in \{1, 2, \ldots, p\};\ \beta_1 \in \{1, 2, \ldots, q\})$ *for the* x_{j_1}-*direction has the following properties.*

- *For* $a_{j_1 j_1 0} \prod_{\alpha_1=1,\alpha_2 \neq \alpha_1}^{p} (a_{j_1 j_1 \alpha_1} - a_{j_1 j_1 \alpha_2})^{m_{\alpha_2}} \prod_{\beta_2=1,\beta_2 \neq \beta_1}^{q} (a_{j_1 j_2 \beta_1} - a_{j_1 j_2 \beta_2})^{n_{\beta_2}} > 0$ *and* $a_{j_2 j_2 0} > 0$,

$$(a_{j_1 j_1 \alpha_1}, a_{j_1 j_2 \beta_1}) = \underbrace{([(2m_{\alpha_1 1}), (2n_{\beta_1 1})]_{\text{II:II}}\text{US,pF})}_{[(2m_{\alpha_1 1}),(2n_{\beta_1 1})]\text{-increasing-inflection upper-saddle flow }(+)}. \quad (1.54)$$

The flow at $(x^*_{j_1}, x^*_{j_2}) = (a_{j_1 j_1 \alpha_1}, a_{j_1 j_2 \beta_1})$ *is a* $([(2m_{\alpha_1 1}), (2n_{\beta_1 1})]_{\text{II:II}}\text{US,pF})$-*positive increasing-inflection upper-saddle flow. The line of* $x^*_{j_1} = a_{j_1 j_1 \alpha_1}$ *is a* $(2m_{\alpha_1 1})^{\text{th}}$-*order positive upper-saddle directrix flow.*

- *For* $a_{j_1 j_1 0} \prod_{\alpha_1=1,\alpha_2 \neq \alpha_1}^{p} (a_{j_1 j_1 \alpha_1} - a_{j_1 j_1 \alpha_2})^{m_{\alpha_2}} \prod_{\beta_2=1,\beta_2 \neq \beta_1}^{q} (a_{j_1 j_2 \beta_1} - a_{j_1 j_2 \beta_2})^{n_{\beta_2}} < 0$ *and* $a_{j_2 j_2 0} > 0$,

$$(a_{j_1 j_1 \alpha_1}, a_{j_1 j_2 \beta_1}) = \underbrace{([(2m_{\alpha_1 1}), (2n_{\beta_1 1})]_{\text{DI:DI}}\text{LS,pF})}_{[(2m_{\alpha_1 1}),(2n_{\beta_1 1})]\text{-decreasing-inflection lower-saddle flow }(+)}. \quad (1.55)$$

The flow at $(x^*_{j_1}, x^*_{j_2}) = (a_{j_1 j_1 \alpha_1}, a_{j_1 j_2 \beta_1})$ *is a* $([(2m_{\alpha_1 1}), (2n_{\beta_1 1})]_{\text{DI:DI}}\text{LS,pF})$-*positive decreasing-inflection lower-saddle flow. The line of* $x^*_{j_1} = a_{j_1 j_1 \alpha_1}$ *is a* $(2m_{\alpha_1 1})^{\text{th}}$-*order positive lower-saddle directrix flow.*

- *For* $a_{j_1 j_1 0} \prod_{\alpha_1=1,\alpha_2 \neq \alpha_1}^{p} (a_{j_1 j_1 \alpha_1} - a_{j_1 j_1 \alpha_2})^{m_{\alpha_2}} \prod_{\beta_2=1,\beta_2 \neq \beta_1}^{q} (a_{j_1 j_2 \beta_1} - a_{j_1 j_2 \beta_2})^{n_{\beta_2}} > 0$ *and* $a_{j_2 j_2 0} < 0$,

$$(a_{j_1 j_1 \alpha_1}, a_{j_1 j_2 \beta_1}) = \underbrace{([(2m_{\alpha_1 1}), (2n_{\beta_1 1})]_{\text{DI:DI}}\text{US,nF})}_{[(2m_{\alpha_1 1}),(2n_{\beta_1 1})]\text{-decreasing-inflection upper-saddle flow }(-)}. \quad (1.56)$$

The flow at $(x^*_{j_1}, x^*_{j_2}) = (a_{j_1j_1\alpha_1}, a_{j_1j_2\beta_1})$ *is a* ($[(2m_{\alpha_1 1}), (2n_{\beta_1 1})]_{\text{DI:DI}}$US,nF)-*negative decreasing-inflection upper-saddle flow. The line of* $x^*_{j_1} = a_{j_1j_1\alpha_1}$ *is a* $(2m_{\alpha_1 1})^{\text{th}}$-order negative upper-saddle *directrix flow.*

- *For* $a_{j_1j_10}\prod_{\alpha_1=1,\alpha_2\neq\alpha_1}^{p}(a_{j_1j_1\alpha_1}-a_{j_1j_1\alpha_2})^{m_{\alpha_2}}\prod_{\beta_2=1,\beta_2\neq\beta_1}^{q}(a_{j_1j_2\beta_1}-a_{j_1j_2\beta_2})^{n_{\beta_2}}<0$ *and* $a_{j_2j_20}<0$,

$$(a_{j_1j_1\alpha_1}, a_{j_1j_2\beta_1}) = \underbrace{([(2m_{\alpha_1 1}),(2n_{\beta_1 1})]_{\text{II:II}}\text{LS,nF})}_{[(2m_{\alpha_1 1}),(2n_{\beta_1 1})]\text{-increasing-inflection lower-saddle flow }(-)}. \tag{1.57}$$

The flow at $(x^*_{j_1}, x^*_{j_2}) = (a_{j_1j_1\alpha_1}, a_{j_1j_2\beta_1})$ *is a* ($[(2m_{\alpha_1 1}), (2n_{\beta_1 1})]_{\text{II:II}}$LS,nF)-*negative increasing-inflection lower-saddle flow. The line of* $x^*_{j_1} = a_{j_1j_1\alpha_1}$ *is a* $(2m_{\alpha_1 1})^{\text{th}}$*-order negative lower-saddle directrix flow.*

(iv) Consider

$$\begin{aligned}\dot{x}_{j_2} &= a_{j_2j_20}\\ \dot{x}_{j_1} &= a_{j_1j_10}\prod_{\alpha=1}^{m}(x_{j_1}-a_{j_1j_1\alpha})\prod_{\beta=1}^{n}(x_{j_2}-a_{j_1j_2\beta}).\end{aligned} \tag{1.58}$$

The first integral manifold is

$$\begin{aligned}&a_{j_2j_20}\sum_{\alpha_1=1}^{m}\frac{1}{\prod_{\alpha_2=1,\alpha_2\neq\alpha_1}^{m}(a_{j_1j_1\alpha_1}-a_{j_1j_1\alpha_2})}\ln\frac{|x_{j_1}-a_{j_1j_1\alpha_1}|}{|x_{j_10}-a_{j_1j_1\alpha_1}|}\\ &=a_{j_1j_10}\{\frac{1}{n+1}\big[(x_{j_2}-a_{j_1j_2\beta_1})^{n+1}-(x_{j_20}-a_{j_1j_2\beta_1})^{n+1}\big]\\ &+\sum_{k=1}^{n-1}\frac{1}{n-k+1}b_{j_1j_2k}\big[(x_{j_2}-a_{j_1j_2\beta_1})^{n-k+1}-(x_{j_20}-a_{j_1j_2\beta_1})^{n-k+1}\big]\}.\end{aligned} \tag{1.59}$$

where

$$\begin{aligned}b_{j_1j_21} &= \sum_{\beta_2=1,\beta_2\neq\beta_1}^{n}(a_{j_1j_2\beta_1}-a_{j_1j_2\beta_2}),\\ b_{j_1j_22} &= \sum_{\substack{\beta_2,\beta_3=1;\beta_2,\beta_3\neq\beta_1\\(\beta_2<\beta_3)}}^{n}\prod_{r=2}^{3}(a_{j_1j_2\beta_1}-a_{j_1j_2\beta_r}),\ldots,\\ b_{j_1j_2k} &= \sum_{\substack{\beta_2,\beta_3,\cdots,\beta_{k+1}=1;\\ \beta_2,\beta_3,\cdots,\beta_{k+1}\neq\beta_1\\(\beta_2<\beta_3<\cdots<\beta_{k+1})}}^{n}\prod_{r=2}^{k+1}(a_{j_1j_2\beta_1}-a_{j_1j_2\beta_r}),\ldots,\\ b_{j_1j_2(m-1)} &= \prod_{\beta_1=1,\beta_2\neq s_1}^{n}(a_{j_1j_2\beta_1}-a_{j_1j_2\beta_2}).\end{aligned} \tag{1.60}$$

(iv_1) The equilibrium of $(x_{j_1}^*, x_{j_2}^*) = (a_{j_1j_1\alpha_1}, a_{j_1j_2\beta_1})$ ($\alpha_1 = 1, 2, \ldots, m$; $\beta_1 = 1, 2, \ldots, n$) for the x_{j_1}-direction has the following properties.

- For $a_{j_1j_10}\prod_{\alpha_2=1,\alpha_2\neq\alpha_1}^{m}(a_{j_1j_1\alpha_1}-a_{j_1j_1\alpha_2})\prod_{\beta_2=1,\beta_2\neq\beta_1}^{n}(a_{j_1j_2\beta_2}-a_{j_1j_2\beta_2}) > 0$ and $a_{j_2j_20} > 0$

$$(a_{j_1j_1\alpha_1}, a_{j_1j_2\beta_1}) = \underbrace{(\text{DP:UP,pF})}_{\text{hyperbolic flow }(+)}. \tag{1.61}$$

- For $a_{j_1j_10}\prod_{\alpha_2=1,\alpha_2\neq\alpha_1}^{m}(a_{j_1j_1\alpha_1}-a_{j_1j_1\alpha_2})\prod_{\beta_2=1,\beta_2\neq\beta_1}^{n}(a_{j_1j_2\beta_2}-a_{j_1j_2\beta_2}) < 0$ and $a_{j_2j_20} > 0$

$$(a_{j_1j_1\alpha_1}, a_{j_1j_2\beta_1}) = \underbrace{(\text{UP:DP,pF})}_{\text{hyperbolic - secant flow }(+)}. \tag{1.62}$$

- For $a_{j_1j_10}\prod_{\alpha_2=1,\alpha_2\neq\alpha_1}^{m}(a_{j_1j_1\alpha_1}-a_{j_1j_1\alpha_2})\prod_{\beta_2=1,\beta_2\neq\beta_1}^{n}(a_{j_1j_2\beta_2}-a_{j_1j_2\beta_2}) > 0$ and $a_{j_2j_20} < 0$

$$(a_{j_1j_1\alpha_1}, a_{j_1j_2\beta_1}) = \underbrace{(\text{UP:DP,nF})}_{\text{hyperbolic - secant flow }(-)}. \tag{1.63}$$

- For $a_{j_1j_10}\prod_{\alpha_2=1,\alpha_2\neq\alpha_1}^{m}(a_{j_1j_1\alpha_1}-a_{j_1j_1\alpha_2})\prod_{\beta_2=1,\beta_2\neq\beta_1}^{n}(a_{j_1j_2\beta_2}-a_{j_1j_2\beta_2}) < 0$ and $a_{j_2j_20} < 0$

$$(a_{j_1j_1\alpha_1}, a_{j_1j_2\beta_1}) = \underbrace{(\text{DP:UP,nF})}_{\text{hyperbolic flow }(-)}. \tag{1.64}$$

(iv_2)There are four cases: *(iii_{2a}) $m = 2m_1 + 1$ and $n = 2n_1 + 1$, (iii_{2b}) $m = 2m_1$ and $n = 2n_1 + 1$, (iii_{2c}) $m = 2m_1 + 1$ and $n = 2n_1$, (iii_{2d}) $m = 2m_1$ and $n = 2n_1$.*

Define

$$\bigcup_{\alpha=1}^{m}\bigcup_{\beta=1}^{n}\underbrace{(a_{j_1j_1\alpha}, a_{j_1j_2\beta})}_{(\text{XX, xF})}$$
$$= \begin{bmatrix} (a_{j_1j_11}, a_{j_1j_2n}) & \cdots & (a_{j_1j_1(m-1)}, a_{j_1j_2n}) & (a_{j_1j_1m}, a_{j_1j_2n}) \\ (a_{j_1j_11}, a_{j_1j_2(n-1)}) & \cdots & (a_{j_1j_1(m-1)}, a_{j_1j_2(n-1)}) & (a_{j_1j_1m}, a_{j_1j_2(n-1)}) \\ \vdots & \vdots & \vdots & \vdots \\ (a_{j_1j_11}, a_{j_1j_21}) & \cdots & (a_{j_1j_1(m-1)}, a_{j_1j_21}) & (a_{j_1j_1m}, a_{j_1j_21}) \end{bmatrix}_{n\times m}. \tag{1.65}$$

(iv_{2a}) For $m = 2m_1+1$ and $n = 2n_1+1$, the equilibriums of $(x^*_{j_1}, x^*_{j_2}) = (a_{j_1j_1\alpha}, a_{j_1j_2\beta})$ $(\alpha \in \{1, 2, \ldots, 2m_1 + 1\},\ \beta \in \{1, 2, \ldots, 2n_1 + 1\})$ *for the x_{j_1}-direction are a hyperbolic and hyperbolic-secant flow network as follows.*

- *For $a_{j_1j_10} > 0$ and $a_{j_2j_20} > 0$,*

$$
\bigcup_{\alpha=1}^{2m_1+1}\bigcup_{\beta=1}^{2n_1+1}\underbrace{(a_{j_1j_1\alpha}, a_{j_1j_2\beta})}_{\text{(XX, pF)}}
= \begin{bmatrix}
\underbrace{\text{(DP:UP,pF)}}_{\text{hyperbolic flow }(+)} & \cdots & \underbrace{\text{(UP:DP,pF)}}_{\text{hyperbolic-secant flow }(+)} & \underbrace{\text{(DP:UP,pF)}}_{\text{hyperbolic flow }(+)} \\
\underbrace{\text{(UP:DP,pF)}}_{\text{hyperbolic-secant flow }(+)} & \cdots & \underbrace{\text{(DP:UP,pF)}}_{\text{hyperbolic flow }(+)} & \underbrace{\text{(UP:DP,pF)}}_{\text{hyperbolic-secant flow }(+)} \\
\vdots & \vdots & \vdots & \vdots \\
\underbrace{\text{(DP:UP,pF)}}_{\text{hyperbolic flow }(+)} & \cdots & \underbrace{\text{(UP:DP,pF)}}_{\text{hyperbolic-secant flow }(+)} & \underbrace{\text{(DP:UP,pF)}}_{\text{hyperbolic flow }(+)}
\end{bmatrix}_{(2n_1+1)\times(2m_1+1)}. \tag{1.66}
$$

- *For $a_{j_1j_10} < 0$ and $a_{j_2j_20} > 0$,*

$$
\bigcup_{\alpha=1}^{2m_1+1}\bigcup_{\beta=1}^{2n_1+1}\underbrace{(a_{j_1j_1\alpha}, a_{j_1j_2\beta})}_{\text{(XX, pF)}}
= \begin{bmatrix}
\underbrace{\text{(UP:DP,pF)}}_{\text{hyperbolic-secant flow }(+)} & \cdots & \underbrace{\text{(DP:UP,pF)}}_{\text{hyperbolic flow }(+)} & \underbrace{\text{(UP:DP,pF)}}_{\text{hyperbolic-secant flow }(+)} \\
\underbrace{\text{(DP:UP,pF)}}_{\text{hyperbolic flow }(+)} & \cdots & \underbrace{\text{(UP:DP,pF)}}_{\text{hyperbolic-secant flow }(+)} & \underbrace{\text{(DP:UP,pF)}}_{\text{hyperbolic flow }(+)} \\
\vdots & \vdots & \vdots & \vdots \\
\underbrace{\text{(UP:DP,pF)}}_{\text{hyperbolic-secant flow }(+)} & \cdots & \underbrace{\text{(DP:UP,pF)}}_{\text{hyperbolic flow }(+)} & \underbrace{\text{(UP:DP,pF)}}_{\text{hyperbolic-secant flow }(+)}
\end{bmatrix}_{(2n_1+1)\times(2m_1+1)}. \tag{1.67}
$$

- *For $a_{j_1j_10} > 0$ and $a_{j_2j_20} < 0$,*

$$
\bigcup_{\alpha=1}^{2m_1+1}\bigcup_{\beta=1}^{2n_1+1}\underbrace{(a_{j_1j_1\alpha}, a_{j_1j_2\beta})}_{\text{(XX, nF)}}
= \begin{bmatrix}
\underbrace{\text{(UP:DP,nF)}}_{\text{hyperbolic-secant flow }(-)} & \cdots & \underbrace{\text{(DP:UP,nF)}}_{\text{hyperbolic flow }(-)} & \underbrace{\text{(UP:DP,nF)}}_{\text{hyperbolic-secant flow }(-)} \\
\underbrace{\text{(DP:UP,nF)}}_{\text{hyperbolic flow }(-)} & \cdots & \underbrace{\text{(UP:DP,nF)}}_{\text{hyperbolic-secant flow }(-)} & \underbrace{\text{(DP:UP,nF)}}_{\text{hyperbolic flow }(-)} \\
\vdots & \vdots & \vdots & \vdots \\
\underbrace{\text{(UP:DP,nF)}}_{\text{hyperbolic-secant flow }(-)} & \cdots & \underbrace{\text{(DP:UP,nF)}}_{\text{hyperbolic flow }(-)} & \underbrace{\text{(UP:DP,nF)}}_{\text{hyperbolic-secant flow }(-)}
\end{bmatrix}_{(2n_1+1)\times(2m_1+1)}. \tag{1.68}
$$

- *For* $a_{j_1j_10} < 0$ *and* $a_{j_2j_20} < 0$,

$$\bigcup_{\alpha=1}^{2m_1+1}\bigcup_{\beta=1}^{2n_1+1}\underbrace{(a_{j_1j_1\alpha},a_{j_1j_2\beta})}_{\text{(XX, nF)}}$$
$$=\begin{bmatrix}\underbrace{\text{(DP:UP,nF)}}_{\text{hyperbolic flow }(-)} & \cdots & \underbrace{\text{(UP:DP,nF)}}_{\text{hyperbolic-secant flow }(-)} & \underbrace{\text{(DP:UP,nF)}}_{\text{hyperbolic flow }(-)}\\ \underbrace{\text{(UP:DP,nF)}}_{\text{hyperbolic-secant flow }(-)} & \cdots & \underbrace{\text{(DP:UP,nF)}}_{\text{hyperbolic flow }(-)} & \underbrace{\text{(UP:DP,nF)}}_{\text{hyperbolic-secant flow }(-)}\\ \vdots & \vdots & \vdots & \vdots\\ \underbrace{\text{(DP:UP,nF)}}_{\text{hyperbolic flow }(-)} & \cdots & \underbrace{\text{(UP:DP,nF)}}_{\text{hyperbolic-secant flow }(-)} & \underbrace{\text{(DP:UP,nF)}}_{\text{hyperbolic flow }(-)}\end{bmatrix}_{(2n_1+1)\times(2m_1+1)}. \tag{1.69}$$

The lines of $x_{j_1}^* = a_{j_1j_1\alpha}$ $(\alpha \in \{1, 2, \ldots, 2m_1 + 1\})$ *are directrix flows.*

(iv$_{2b}$*) For* $m = 2m_1$ *and* $n = 2n_1 + 1$, *the equilibriums of* $(x_{j_1}^*, x_{j_2}^*) = (a_{j_1j_1\alpha}, a_{j_1j_2\beta})$ $(\alpha \in \{1, 2, \ldots, 2m_1\}, \beta \in \{1, 2, \ldots, 2n_1 + 1\})$ *for the* x_{j_1}*-direction are a hyperbolic and hyperbolic-secant flow network as follows.*

- *For* $a_{j_1j_10} > 0$ *and* $a_{j_2j_20} > 0$,

$$\bigcup_{\alpha=1}^{2m_1}\bigcup_{\beta=1}^{2n_1+1}\underbrace{(a_{j_1j_1\alpha},a_{j_1j_2\beta})}_{\text{(XX, pF)}}$$
$$=\begin{bmatrix}\underbrace{\text{(UP:DP,pF)}}_{\text{hyperbolic-secant flow }(+)} & \cdots & \underbrace{\text{(UP:DP,pF)}}_{\text{hyperbolic-secant flow }(+)} & \underbrace{\text{(DP:UP,pF)}}_{\text{hyperbolic flow }(+)}\\ \underbrace{\text{(DP:UP,pF)}}_{\text{hyperbolic flow }(+)} & \cdots & \underbrace{\text{(DP:UP,pF)}}_{\text{hyperbolic flow }(+)} & \underbrace{\text{(UP:DP,pF)}}_{\text{hyperbolic-secant flow }(+)}\\ \vdots & \vdots & \vdots & \vdots\\ \underbrace{\text{(UP:DP,pF)}}_{\text{hyperbolic-secant flow }(+)} & \cdots & \underbrace{\text{(UP:DP,pF)}}_{\text{hyperbolic-secant flow }(+)} & \underbrace{\text{(DP:UP,pF)}}_{\text{hyperbolic flow }(+)}\end{bmatrix}_{(2n_1+1)\times(2m_1)}. \tag{1.70}$$

- *For* $a_{j_1j_10} < 0$ *and* $a_{j_2j_20} > 0$,

$$\bigcup_{\alpha=1}^{2m_1}\bigcup_{\beta=1}^{2n_1+1}\underbrace{(a_{j_1j_1\alpha},a_{j_1j_2\beta})}_{\text{(XX, pF)}}$$
$$=\begin{bmatrix}\underbrace{\text{(DP:UP,pF)}}_{\text{hyperbolic flow }(+)} & \cdots & \underbrace{\text{(DP:UP,pF)}}_{\text{hyperbolic flow }(+)} & \underbrace{\text{(UP:DP,pF)}}_{\text{hyperbolic-secant flow }(+)}\\ \underbrace{\text{(UP:DP,pF)}}_{\text{hyperbolic-secant flow }(+)} & \cdots & \underbrace{\text{(UP:DP,pF)}}_{\text{hyperbolic-secant flow }(+)} & \underbrace{\text{(DP:UP,pF)}}_{\text{hyperbolic flow }(+)}\\ \vdots & \vdots & \vdots & \vdots\\ \underbrace{\text{(DP:UP,pF)}}_{\text{hyperbolic flow }(+)} & \cdots & \underbrace{\text{(DP:UP,pF)}}_{\text{hyperbolic flow }(+)} & \underbrace{\text{(UP:DP,pF)}}_{\text{hyperbolic-secant flow }(+)}\end{bmatrix}_{(2n_1+1)\times(2m_1)}. \tag{1.71}$$

- *For $a_{j_1j_10} > 0$ and $a_{j_2j_20} < 0$,*

$$\bigcup_{\alpha=1}^{2m_1}\bigcup_{\beta=1}^{2n_1+1}\underbrace{(a_{j_1j_1\alpha},a_{j_1j_2\beta})}_{\text{(XX, nF)}}$$

$$=\begin{bmatrix}\underbrace{\text{(DP:UP,nF)}}_{\text{hyperbolic flow }(-)} & \cdots & \underbrace{\text{(DP:UP,nF)}}_{\text{hyperbolic flow }(-)} & \underbrace{\text{(UP:DP,nF)}}_{\text{hyperbolic-secant flow }(-)}\\ \underbrace{(UP:DP,nF)}_{\text{hyperbolic-secant flow }(-)} & \cdots & \underbrace{\text{(UP:DP,nF)}}_{\text{hyperbolic-secant flow }(-)} & \underbrace{\text{(DP:UP,nF)}}_{\text{hyperbolic flow }(-)}\\ \vdots & \vdots & \vdots & \vdots\\ \underbrace{\text{(DP:UP,nF)}}_{\text{hyperbolic flow }(-)} & \cdots & \underbrace{\text{(DP:UP,nF)}}_{\text{hyperbolic flow }(-)} & \underbrace{\text{(UP:DP,nF)}}_{\text{hyperbolic-secant flow }(-)}\end{bmatrix}_{(2n_1+1)\times(2m_1)}. \tag{1.72}$$

- *For $a_{j_1j_10} < 0$ and $a_{j_2j_20} < 0$,*

$$\bigcup_{\alpha=1}^{2m_1}\bigcup_{\beta=1}^{2n_1+1}\underbrace{(a_{j_1j_1\alpha},a_{j_1j_2\beta})}_{\text{(XX, nF)}}$$

$$=\begin{bmatrix}\underbrace{\text{(UP:DP,nF)}}_{\text{hyperbolic-secant flow }(-)} & \cdots & \underbrace{\text{(UP:DP,nF)}}_{\text{hyperbolic-secant flow }(-)} & \underbrace{\text{(DP:UP,nF)}}_{\text{hyperbolic flow }(-)}\\ \underbrace{\text{(DP:UP,nF)}}_{\text{hyperbolic flow }(-)} & \cdots & \underbrace{\text{(DP:UP,nF)}}_{\text{hyperbolic flow }(-)} & \underbrace{\text{(UP:DP,nF)}}_{\text{hyperbolic-secant flow }(-)}\\ \vdots & \vdots & \vdots & \vdots\\ \underbrace{\text{(UP:DP,nF)}}_{\text{hyperbolic-secant flow }(-)} & \cdots & \underbrace{\text{(UP:DP,nF)}}_{\text{hyperbolic-secant flow }(-)} & \underbrace{\text{(DP:UP,nF)}}_{\text{hyperbolic flow }(-)}\end{bmatrix}_{(2n_1+1)\times(2m_1)}. \tag{1.73}$$

The lines of $x_{j_1}^* = a_{j_1j_1\alpha}$ ($\alpha \in \{1,2,\ldots,2m_1\}$) are directrix flows.

(iv_{2c}) For $m = 2m_1 + 1$ and $n = 2n_1$, the equilibriums of $(x_{j_1}^*, x_{j_2}^*) = (a_{j_1j_1\alpha}, a_{j_1j_2\beta})$ ($\alpha \in \{1,2,\ldots,2m_1+1\}, \beta \in \{1,2,\ldots,2n_1\}$) for the x_{j_1}-direction for the x_{j_1}-direction are a hyperbolic and hyperbolic-secant flow network as follows.

- *For $a_{j_1j_10} > 0$ and $a_{j_2j_20} > 0$,*

$$\bigcup_{\alpha=1}^{2m_1+1}\bigcup_{\beta=1}^{2n_1}\underbrace{(a_{j_1j_1\alpha},a_{j_1j_2\beta})}_{\text{(XX, pF)}}$$

$$=\begin{bmatrix}\underbrace{\text{(DP:UP,pF)}}_{\text{hyperbolic flow }(+)} & \cdots & \underbrace{\text{(UP:DP,pF)}}_{\text{hyperbolic-secant flow }(+)} & \underbrace{\text{(DP:UP,pF)}}_{\text{hyperbolic flow }(+)}\\ \underbrace{\text{(UP:DP,pF)}}_{\text{hyperbolic-secant flow }(+)} & \cdots & \underbrace{\text{(DP:UP,pF)}}_{\text{hyperbolic flow }(+)} & \underbrace{\text{(UP:DP,pF)}}_{\text{hyperbolic-secant flow }(+)}\\ \vdots & \vdots & \vdots & \vdots\\ \underbrace{\text{(UP:DP,pF)}}_{\text{hyperbolic-secant flow }(+)} & \cdots & \underbrace{\text{(DP:UP,pF)}}_{\text{hyperbolic flow }(+)} & \underbrace{\text{(UP:DP,pF)}}_{\text{hyperbolic-secant flow }(+)}\end{bmatrix}_{(2n_1)\times(2m_1+1)}. \tag{1.74}$$

- *For* $a_{j_1j_10} < 0$ *and* $a_{j_2j_20} > 0$,

$$\bigcup_{\alpha=1}^{2m_1+1}\bigcup_{\beta=1}^{2n_1}\underbrace{(a_{j_1j_1\alpha},a_{j_1j_2\beta})}_{\text{(XX, pF)}} = \begin{bmatrix} \underbrace{\text{(UP:DP,pF)}}_{\text{hyperbolic-secant flow }(+)} & \cdots & \underbrace{\text{(DP:UP,pF)}}_{\text{hyperbolic flow }(+)} & \underbrace{\text{(UP:DP,pF)}}_{\text{hyperbolic-secant flow }(+)} \\ \underbrace{\text{(DP:UP,pF)}}_{\text{hyperbolic flow }(+)} & \cdots & \underbrace{\text{(UP:DP,pF)}}_{\text{hyperbolic-secant flow }(+)} & \underbrace{\text{(DP:UP,pF)}}_{\text{hyperbolic flow }(+)} \\ \vdots & \vdots & \vdots & \vdots \\ \underbrace{\text{(DP:UP,pF)}}_{\text{hyperbolic flow }(+)} & \cdots & \underbrace{\text{(UP:DP,pF)}}_{\text{hyperbolic-secant flow }(+)} & \underbrace{\text{(DP:UP,pF)}}_{\text{hyperbolic flow }(+)} \end{bmatrix}_{(2n_1)\times(2m_1+1)}. \tag{1.75}$$

- *For* $a_{j_1j_10} > 0$ *and* $a_{j_2j_20} < 0$,

$$\bigcup_{\alpha=1}^{2m_1+1}\bigcup_{\beta=1}^{2n_1}\underbrace{(a_{j_1j_1\alpha},a_{j_1j_2\beta})}_{\text{(XX, nF)}} = \begin{bmatrix} \underbrace{\text{(UP:DP,nF)}}_{\text{hyperbolic-secant flow }(-)} & \cdots & \underbrace{\text{(DP:UP,nF)}}_{\text{hyperbolic flow }(-)} & \underbrace{\text{(UP:DP,nF)}}_{\text{hyperbolic-secant flow }(-)} \\ \underbrace{\text{(DP:UP,nF)}}_{\text{hyperbolic flow }(-)} & \cdots & \underbrace{\text{(UP:DP,nF)}}_{\text{hyperbolic-secant flow }(-)} & \underbrace{\text{(DP:UP,nF)}}_{\text{hyperbolic flow }(-)} \\ \vdots & \vdots & \vdots & \vdots \\ \underbrace{\text{(DP:UP,nF)}}_{\text{hyperbolic flow }(-)} & \cdots & \underbrace{\text{(UP:DP,nF)}}_{\text{hyperbolic-secant flow }(-)} & \underbrace{\text{(DP:UP,nF)}}_{\text{hyperbolic flow }(-)} \end{bmatrix}_{(2n_1)\times(2m_1+1)}. \tag{1.76}$$

- *For* $a_{j_1j_10} < 0$ *and* $a_{j_2j_20} < 0$,

$$\bigcup_{\alpha=1}^{2m_1+1}\bigcup_{\beta=1}^{2n_1}\underbrace{(a_{j_1j_1\alpha},a_{j_1j_2\beta})}_{\text{(XX, nF)}} = \begin{bmatrix} \underbrace{\text{(DP:UP,nF)}}_{\text{hyperbolic flow }(-)} & \cdots & \underbrace{\text{(UP:DP,nF)}}_{\text{hyperbolic-secant flow }(-)} & \underbrace{\text{(DP:UP,nF)}}_{\text{hyperbolic flow }(-)} \\ \underbrace{\text{(UP:DP,nF)}}_{\text{hyperbolic-secant flow }(-)} & \cdots & \underbrace{\text{(DP:UP,nF)}}_{\text{hyperbolic flow }(-)} & \underbrace{\text{(UP:DP,nF)}}_{\text{hyperbolic-secant flow }(-)} \\ \vdots & \vdots & \vdots & \vdots \\ \underbrace{\text{(UP:DP,nF)}}_{\text{hyperbolic-secant flow }(-)} & \cdots & \underbrace{\text{(DP:UP,nF)}}_{\text{hyperbolic flow }(-)} & \underbrace{\text{(UP:DP,nF)}}_{\text{hyperbolic-secant flow }(-)} \end{bmatrix}_{(2n_1)\times(2m_1+1)}. \tag{1.77}$$

The lines of $x_{j_1}^* = a_{j_1j_1\alpha}$ $(\alpha \in \{1, 2, \ldots, 2m_1 + 1\})$ *are directrix flows.*

(iv_{2d}) For $m = 2m_1$ and $n = 2n_1$, the equilibriums of $(x^*_{j_1}, x^*_{j_2}) = (a_{j_1j_1\alpha}, a_{j_1j_2\beta})$ ($\alpha \in \{1, 2, \ldots, 2m_1\}, \beta \in \{1, 2, \ldots, 2n_1\}$) for the x_{j_1}-direction are a hyperbolic and hyperbolic-secant flow network as follows.

- *For $a_{j_1j_10} > 0$ and $a_{j_2j_20} > 0$,*

$$\bigcup_{\alpha=1}^{2m_1}\bigcup_{\beta=1}^{2n_1}\underbrace{(a_{j_1j_1\alpha}, a_{j_1j_2\beta})}_{\text{(XX, pF)}} = \begin{bmatrix} \underbrace{\text{(UP:DP,pF)}}_{\text{hyperbolic-secant flow }(+)} & \cdots & \underbrace{\text{(UP:DP,pF)}}_{\text{hyperbolic-secant flow }(+)} & \underbrace{\text{(DP:UP,pF)}}_{\text{hyperbolic flow }(+)} \\ \underbrace{\text{(DP:UP,pF)}}_{\text{hyperbolic flow }(+)} & \cdots & \underbrace{\text{(DP:UP,pF)}}_{\text{hyperbolic flow }(+)} & \underbrace{\text{(UP:DP,pF)}}_{\text{hyperbolic-secant flow }(+)} \\ \vdots & \vdots & \vdots & \vdots \\ \underbrace{\text{(DP:UP,pF)}}_{\text{hyperbolic flow }(+)} & \cdots & \underbrace{\text{(DP:UP,pF)}}_{\text{hyperbolic flow }(+)} & \underbrace{\text{(UP:DP,pF)}}_{\text{hyperbolic-secant flow }(+)} \end{bmatrix}_{(2n_1)\times(2m_1)}. \tag{1.78}$$

- *For $a_{j_1j_10} < 0$ and $a_{j_2j_20} > 0$,*

$$\bigcup_{\alpha=1}^{2m_1}\bigcup_{\beta=1}^{2n_1}\underbrace{(a_{j_1j_1\alpha}, a_{j_1j_2\beta})}_{\text{(XX, pF)}} = \begin{bmatrix} \underbrace{\text{(DP:UP,pF)}}_{\text{hyperbolic flow }(+)} & \cdots & \underbrace{\text{(DP:UP,pF)}}_{\text{hyperbolic flow }(+)} & \underbrace{\text{(UP:DP,pF)}}_{\text{hyperbolic-secant flow }(+)} \\ \underbrace{\text{(UP:DP,pF)}}_{\text{hyperbolic-secant flow }(+)} & \cdots & \underbrace{\text{(UP:DP,pF)}}_{\text{hyperbolic-secant flow }(+)} & \underbrace{\text{(DP:UP,pF)}}_{\text{hyperbolic flow }(+)} \\ \vdots & \vdots & \vdots & \vdots \\ \underbrace{\text{(UP:DP,pF)}}_{\text{hyperbolic-secant flow }(+)} & \cdots & \underbrace{\text{(UP:DP,pF)}}_{\text{hyperbolic-secant flow }(+)} & \underbrace{\text{(DP:UP,pF)}}_{\text{hyperbolic flow }(+)} \end{bmatrix}_{(2n_1)\times(2m_1)}. \tag{1.79}$$

- *For $a_{j_1j_10} > 0$ and $a_{j_2j_20} < 0$,*

$$\bigcup_{\alpha=1}^{2m_1}\bigcup_{\beta=1}^{2n_1}\underbrace{(a_{j_1j_1\alpha}, a_{j_1j_2\beta})}_{\text{(XX, nF)}} = \begin{bmatrix} \underbrace{\text{(DP:UP,nF)}}_{\text{hyperbolic flow }(-)} & \cdots & \underbrace{\text{(DP:UP,nF)}}_{\text{hyperbolic flow }(-)} & \underbrace{\text{(UP:DP,nF)}}_{\text{hyperbolic-secant flow }(-)} \\ \underbrace{\text{(UP:DP,nF)}}_{\text{hyperbolic-secant flow }(-)} & \cdots & \underbrace{\text{(UP:DP,nF)}}_{\text{hyperbolic - secant flow }(-)} & \underbrace{\text{(DP:UP,nF)}}_{\text{hyperbolic flow }(-)} \\ \vdots & \vdots & \vdots & \vdots \\ \underbrace{\text{(UP:DP,nF)}}_{\text{hyperbolic-secant flow }(-)} & \cdots & \underbrace{\text{(UP:DP,nF)}}_{\text{hyperbolic-secant flow }(-)} & \underbrace{\text{(DP:UP,nF)}}_{\text{hyperbolic flow }(-)} \end{bmatrix}_{(2n_1)\times(2m_1)}. \tag{1.80}$$

- *For* $a_{j_1j_10}<0$ *and* $a_{j_2j_20}<0$,

$$\bigcup_{\alpha=1}^{2m_1}\bigcup_{\beta=1}^{2n_1}\underbrace{(a_{j_1j_1\alpha},a_{j_1j_2\beta})}_{\text{(XX, nF)}}$$

$$=\begin{bmatrix}\underbrace{\text{(UP:DP,nF)}}_{\text{hyperbolic-secant flow }(-)} & \cdots & \underbrace{\text{(UP:DP,nF)}}_{\text{hyperbolic-secant flow }(-)} & \underbrace{\text{(DP:UP,nF)}}_{\text{hyperbolic flow }(-)}\\ \underbrace{\text{(DP:UP,nF)}}_{\text{hyperbolic flow }(-)} & \cdots & \underbrace{\text{(DP:UP,nF)}}_{\text{hyperbolic flow }(-)} & \underbrace{\text{(UP:DP,nF)}}_{\text{hyperbolic-secant flow }(-)}\\ \vdots & \vdots & \vdots & \vdots\\ \underbrace{\text{(DP:UP,nF)}}_{\text{hyperbolic flow }(-)} & \cdots & \underbrace{\text{(DP:UP,nF)}}_{\text{hyperbolic flow }(-)} & \underbrace{\text{(UP:DP,nF)}}_{\text{hyperbolic-secant flow }(-)}\end{bmatrix}_{(2n_1)\times(2m_1)}. \tag{1.81}$$

The lines of $x_{j_1}^*=a_{j_1j_1\alpha}$ $(\alpha\in\{1,2,\ldots,2m_1\})$ *are directrix flows.*

(v) For $m=2m_1+1$ *and* $n=2n_1+1$*, the equilibrium of* $(x_{j_1}^*,x_{j_2}^*)=(a_{j_1j_11},a_{j_1j_21})$ *for the* x_{j_1}*-direction has the following bifurcation properties with*

$$\begin{aligned}&2m_1+1-\sum\nolimits_{\alpha_1=1}^{p_1}m_{\alpha_1}=2\sum\nolimits_{\rho_1=1}^{l_1}m_{\rho_1},\;\;2n_1+1-\sum\nolimits_{\beta_1=1}^{q_1}n_{\beta_1}=2\sum\nolimits_{\sigma_1=1}^{l_2}n_{\sigma_1};\\&\sum\nolimits_{\alpha_2=1}^{p_2}m_{\alpha_2}=2m_1+1,\;\;\sum\nolimits_{\beta_2=1}^{q_2}n_{\beta_2}=2n_1+1;\\&\sum\nolimits_{\alpha_i=1}^{p}m_{\alpha_i}=2m_1+1,\;\;\sum\nolimits_{\beta_i=1}^{q}n_{\beta_i}=2n_1+1,\;\text{for } i=1,2;\end{aligned}\tag{1.82}$$

l_1,l_2*-quadratic polynomials are without real roots, and the* ρ_1^{th} *and* σ_1^{th}*-quadratic polynomials are with powers* m_{ρ_1} *and* n_{σ_1}*, respectively.*

(v_1) For $a_{j_1j_10}>0$ *and* $a_{j_2j_20}>0$*, there is a* ([(2m_1+1), (2n_1 + 1)]DP:UP,pF)*-positive hyperbolic flow as*

$$(a_{j_1j_11},a_{j_1j_21})=\underbrace{([(2m_1{+}1),(2n_1+1)]\text{DP:UP,pF}),}_{[(2m_1+1),(2n_1+1)]\text{-hyperbolic flow }(+)}\tag{1.83}$$

there are three following ([(2m_1+1), (2n_1 + 1)]DP:UP,pF)*-positive hyperbolic flow appearing and switching bifurcations.*

(v_{1a}) The ([(2m_1+1), (2n_1 + 1)]DP:UP,pF)*-positive hyperbolic flow appearing bifurcation is from a* (DP:UP,pF)*-positive hyperbolic flow to a hyperbolic and hyperbolic-secant flow network, i.e.,*

$$\underbrace{(a_{j_1j_11}, a_{j_1j_21})}_{\text{(DP:UP, pF)-hyperoblic flow}} \rightleftharpoons \underbrace{([(2m_1+1), (2n_1+1)]\text{DP:UP,pF})}_{[(2m_1+1),(2n_1+1)]\text{-hyperbolic flow }(+)}$$
$$\rightleftharpoons \bigcup\nolimits_{\alpha=1}^{2m_1+1} \bigcup\nolimits_{\beta=1}^{2n_1+1} \underbrace{(a_{j_1j_1\alpha}, a_{j_1j_2\beta})}_{\text{(XX, pF)}}. \tag{1.84}$$

(v_{1b}) The ([(2m_1+1), (2n_1 + 1)]DP:UP,pF)*-positive hyperbolic flow appearing bifurcation is from a $q_1 \times p_1$ to $q_2 \times p_2$ singular hyperbolic and hyperbolic-secant flow networks, i.e.,*

$$\bigcup\nolimits_{\alpha_1=1}^{p_1} \bigcup\nolimits_{\beta_1=1}^{q_1} \underbrace{(a_{j_1j_1\alpha_1}, a_{j_1j_2\beta_1})}_{([m_{\alpha_1}, n_{\beta_1}]\text{XX, pF})} \rightleftharpoons \underbrace{([(2m_1+1), (2n_1+1)]\text{DP:UP,pF})}_{[(2m_1+1),(2n_1+1)]\text{-hyperbolic flow }(+)}$$
$$\rightleftharpoons \bigcup\nolimits_{\alpha_2=1}^{p_2} \bigcup\nolimits_{\beta_2=1}^{q_2} \underbrace{(a_{j_1j_1\alpha_2}, a_{j_1j_2\beta_2})}_{([m_{\alpha_2}, n_{\beta_2}]\text{XX, pF})}. \tag{1.85}$$

(v_{1c}) The ([(2m_1+1), (2n_1 + 1)]DP:UP,pF)*-positive hyperbolic flow switching bifurcation is for two $q \times p$ singular hyperbolic and hyperbolic-secant flow networks, i.e.,*

$$\bigcup\nolimits_{\alpha_1=1}^{p} \bigcup\nolimits_{\beta_1=1}^{q} \underbrace{(a_{j_1j_1\alpha_1}, a_{j_1j_2\beta_1})}_{([m_{\alpha_1}, n_{\beta_1}]\text{XX, pF})} \rightleftharpoons \underbrace{([(2m_1+1), (2n_1+1)]\text{DP:UP,pF})}_{[(2m_1+1),(2n_1+1)]\text{-hyperbolic flow }(+)}$$
$$\rightleftharpoons \bigcup\nolimits_{\alpha_2=1}^{p} \bigcup\nolimits_{\beta_2=1}^{q} \underbrace{(a_{j_1j_1\alpha_2}, a_{j_1j_2\beta_2})}_{([m_{\alpha_2}, n_{\beta_2}]\text{XX, pF})}. \tag{1.86}$$

(v_2) For $a_{j_1j_10} < 0$ and $a_{j_2j_20} > 0$, there is a ([(2m_1+1), (2n_1 + 1)]UP:DP,pF)*-positive hyperbolic-secant flow as*

$$(a_{j_1j_11}, a_{j_1j_21}) = \underbrace{([(2m_1+1), (2n_1+1)]\text{UP:DP,pF})}_{[(2m_1+1),(2n_1+1)]\text{-hyperbolic-secant flow }(+)}, \tag{1.87}$$

there are three following ([(2m_1+1), (2n_1 + 1)]UP:DP,pF)*-positive hyperbolic-secant flow appearing and switching bifurcations.*

(v_{2a}) The ([(2m_1+1), (2n_1+1)]UP:DP,pF)*-positive hyperbolic-secant flow appearing bifurcation is from a* (UP:DP,pF)*-positive hyperbolic-secant flow to a hyperbolic and hyperbolic-secant flow network.*

$$\underbrace{(a_{j_1j_1 1}, a_{j_1j_2 1})}_{\text{(UP:DP, pF)-hyperoblic-secant flow}} \rightleftharpoons \underbrace{([(2m_1{+}1),\,(2n_1+1)]\text{UP:DP,pF})}_{[(2m_1+1),(2n_1+1)]\text{-hyperbolic-secant flow }(+)} \rightleftharpoons \bigcup\nolimits_{\alpha=1}^{2m_1+1} \bigcup\nolimits_{\beta=1}^{2n_1+1} \underbrace{(a_{j_1j_1\alpha}, a_{j_1j_2\beta})}_{\text{(XX, pF)}}. \tag{1.88}$$

(v_{2b}) The ([(2m_1+1), (2n_1 + 1)]UP:DP,pF)*-positive hyperbolic-secant flow appearing bifurcation is from a* $q_1 \times p_1$ *to* $q_2 \times p_2$ *singular hyperbolic and hyperbolic-secant flow networks*

$$\bigcup\nolimits_{\alpha_1=1}^{p_1} \bigcup\nolimits_{\beta_1=1}^{q_1} \underbrace{(a_{j_1j_1\alpha_1}, a_{j_1j_2\beta_1})}_{([m_{\alpha_1}, n_{\beta_1}]\text{XX, pF})} \rightleftharpoons \underbrace{([(2m_1{+}1),\,(2n_1+1)]\text{UP:DP,pF})}_{[(2m_1+1),(2n_1+1)]\text{-hyperbolic-secant flow }(+)} \rightleftharpoons \bigcup\nolimits_{\alpha_2=1}^{p_2} \bigcup\nolimits_{\beta_2=1}^{q_2} \underbrace{(a_{j_1j_1\alpha_2}, a_{j_1j_2\beta_2})}_{([m_{\alpha_2}, n_{\beta_2}]\text{XX, pF})}. \tag{1.89}$$

(v_{2c}) The ([(2m_1+1), (2n_1 + 1)]UP:DP,pF)*-positive hyperbolic flow switching bifurcation is for two* $q \times p$ *singular hyperbolic and hyperbolic-secant flow networks*

$$\bigcup\nolimits_{\alpha_1=1}^{p} \bigcup\nolimits_{\beta_1=1}^{q} \underbrace{(a_{j_1j_1\alpha_1}, a_{j_1j_2\beta_1})}_{([m_{\alpha_1}, n_{\beta_1}]\text{XX, pF})} \rightleftharpoons \underbrace{([(2m_1{+}1),\,(2n_1+1)]\text{UP:DP,pF})}_{[(2m_1+1),(2n_1+1)]\text{-hyperbolic-secant flow }(+)} \rightleftharpoons \bigcup\nolimits_{\alpha_2=1}^{p} \bigcup\nolimits_{\beta_2=1}^{q} \underbrace{(a_{j_1j_1\alpha_2}, a_{j_1j_2\beta_2})}_{([m_{\alpha_2}, n_{\beta_2}]\text{XX, pF})}. \tag{1.90}$$

(v_3) For $a_{j_1j_1 0} > 0$ *and* $a_{j_2j_2 0} < 0$*, there is a* ([(2m_1+1), (2n_1 + 1)]UP:DP,nF)*-negative hyperbolic-secant flow as*

$$(a_{j_1j_1 1}, a_{j_1j_2 1}) = \underbrace{([(2m_1{+}1),\,(2n_1+1)]\text{UP:DP,nF})}_{[(2m_1+1),(2n_1+1)]\text{-hyperbolic-secant flow }(-)}, \tag{1.91}$$

there are three following ([(2m_1+1), (2n_1 + 1)]UP:DP,nF)*-negative hyperbolic-secant flow appearing and switching bifurcations.*

(v_{3a}) The ([(2m_1+1), (2n_1 + 1)]UP:DP,nF)*-negative hyperbolic-secant flow appearing bifurcation is from a* (UP:DP,nF)*-negative hyperbolic-secant flow to a hyperbolic and hyperbolic-secant flow network.*

$$\underbrace{(a_{j_1j_1 1}, a_{j_1j_2 1})}_{\text{(UP:DP, nF)-hyperoblic-secant flow}} \rightleftharpoons \underbrace{([(2m_1{+}1),\,(2n_1+1)]\text{UP:DP,nF})}_{[(2m_1+1),(2n_1+1)]\text{-hyperbolic-secant flow }(-)} \rightleftharpoons \bigcup\nolimits_{\alpha=1}^{2m_1+1} \bigcup\nolimits_{\beta=1}^{2n_1+1} \underbrace{(a_{j_1j_1\alpha}, a_{j_1j_2\beta})}_{\text{(XX, nF)}}. \tag{1.92}$$

(v_{3b}) The ([(2m_1+1), (2n_1+1)]UP:DP,nF)*-negative hyperbolic-secant flow appearing bifurcation is from a $q_1 \times p_1$ to $q_2 \times p_2$ singular hyperbolic and hyperbolic-secant flow networks*

$$\bigcup_{\alpha_1=1}^{p_1}\bigcup_{\beta_1=1}^{q_1}\underbrace{(a_{j_1j_1\alpha_1},a_{j_1j_2\beta_1})}_{([m_{\alpha_1},n_{\beta_1}]\text{XX},\text{nF})}\rightleftharpoons\underbrace{([(2m_1+1),(2n_1+1)]\text{UP:DP,nF})}_{[(2m_1+1),(2n_1+1)]\text{-hyperbolic secant flow }(\ )}$$
$$\rightleftharpoons\bigcup_{\alpha_2=1}^{p_2}\bigcup_{\beta_2=1}^{q_2}\underbrace{(a_{j_1j_1\alpha_2},a_{j_1j_2\beta_2})}_{([m_{\alpha_2},n_{\beta_2}]\text{XX},\text{nF})}. \tag{1.93}$$

(v_{3c}) The ([(2m_1+1), (2n_1 + 1)]UP:DP,pF)*-positive hyperbolic flow switching bifurcation is for two $q \times p$ singular hyperbolic and hyperbolic-secant flow networks*

$$\bigcup_{\alpha_1=1}^{p}\bigcup_{\beta_1=1}^{q}\underbrace{(a_{j_1j_1\alpha_1},a_{j_1j_2\beta_1})}_{([m_{\alpha_1},n_{\beta_1}]\text{XX},\text{nF})}\rightleftharpoons\underbrace{([(2m_1+1),(2n_1+1)]\text{UP:DP,nF})}_{[(2m_1+1),(2n_1+1)]\text{-hyperbolic-secant flow }(-)}$$
$$\rightleftharpoons\bigcup_{\alpha_2=1}^{p}\bigcup_{\beta_2=1}^{q}\underbrace{(a_{j_1j_1\alpha_2},a_{j_1j_2\beta_2})}_{([m_{\alpha_2},n_{\beta_2}]\text{XX},\text{nF})}. \tag{1.94}$$

(v_4) For $a_{j_1j_10} < 0$ and $a_{j_2j_20} < 0$, there is a ([(2m_1+1), (2n_1 + 1)]DP:UP,nF)*-negative hyperbolic flow as*

$$(a_{j_1j_11},a_{j_1j_21})=\underbrace{([(2m_1+1),(2n_1+1)]\text{DP:UP,nF})}_{[(2m_1+1),(2n_1+1)]\text{-hyperbolic flow }(-)}, \tag{1.95}$$

there are three following ([(2m_1+1), (2n_1 + 1)]DP:UP,pF)*-negative hyperbolic flow appearing and switching bifurcations.*

(v_{4a}) The ([(2m_1+1), (2n_1 + 1)]DP:UP,nF)*-negative hyperbolic flow appearing bifurcation is from a* (DP:UP,nF)*-negative hyperbolic flow to a hyperbolic and hyperbolic-secant flow network, i.e.,*

$$\underbrace{(a_{j_1j_11},a_{j_1j_21})}_{(\text{DP:UP, nF})\text{-hyperoblic flow}}\rightleftharpoons\underbrace{([(2m_1+1),(2n_1+1)]\text{DP:UP,nF})}_{[(2m_1+1),(2n_1+1)]\text{-hyperbolic flow }(-)}$$
$$\rightleftharpoons\bigcup_{\alpha=1}^{2m_1+1}\bigcup_{\beta=1}^{2n_1+1}\underbrace{(a_{j_1j_1\alpha},a_{j_1j_2\beta})}_{(\text{XX},\text{nF})}. \tag{1.96}$$

(v_{4b}) The ([(2m_1+1), (2n_1 + 1)]DP:UP,nF)*-negative hyperbolic flow appearing bifurcation is from a $q_1 \times p_1$ to $q_2 \times p_2$ singular hyperbolic and hyperbolic-secant flow networks, i.e.,*

$$\bigcup_{\alpha_1=1}^{p_1}\bigcup_{\beta_1=1}^{q_1}\underbrace{(a_{j_1j_1\alpha_1},a_{j_1j_2\beta_1})}_{([m_{\alpha_1},n_{\beta_1}]\text{XX, nF})}\rightleftharpoons\underbrace{([(2m_1+1),(2n_1+1)]\text{DP:UP,nF})}_{[(2m_1+1),(2n_1+1)]\text{-hyperbolic flow }(-)}$$
$$\rightleftharpoons\bigcup_{\alpha_2=1}^{p_2}\bigcup_{\beta_2=1}^{q_2}\underbrace{(a_{j_1j_1\alpha_2},a_{j_1j_2\beta_2})}_{([m_{\alpha_2},n_{\beta_2}]\text{XX, nF})}. \tag{1.97}$$

(v_{4c}) The ([$(2m_1+1)$, $(2n_1+1)$]DP:UP,nF)-*negative hyperbolic flow switching bifurcation is for two $q \times p$ singular hyperbolic and hyperbolic-secant flow networks, i.e.,*

$$\bigcup_{\alpha_1=1}^{p}\bigcup_{\beta_1=1}^{q}\underbrace{(a_{j_1j_1\alpha_1},a_{j_1j_2\beta_1})}_{([m_{\alpha_1},n_{\beta_1}]\text{XX, nF})}\rightleftharpoons\underbrace{([(2m_1+1),(2n_1+1)]\text{DP:UP,pF})}_{[(2m_1+1),(2n_1+1)]\text{-hyperbolic flow }(+)}$$
$$\rightleftharpoons\bigcup_{\alpha_2=1}^{p}\bigcup_{\beta_2=1}^{q}\underbrace{(a_{j_1j_1\alpha_2},a_{j_1j_2\beta_2})}_{([m_{\alpha_2},n_{\beta_2}]\text{XX, nF})}. \tag{1.98}$$

(vi) For $m = 2m_1$ and $n = 2n_1 + 1$, the equilibrium of $(x_{j_1}^*, x_{j_2}^*) = (a_{j_1j_1 1}, a_{j_1j_2 1})$ for the x_{j_1}-direction has the following bifurcation properties with

$$\begin{aligned}
&2m_1-\sum_{\alpha_1=1}^{p_1}m_{\alpha_1}=2\sum_{\rho_1=1}^{l_1}m_{\rho_1},\ 2n_1+1-\sum_{\beta_1=1}^{q_1}n_{\beta_1}=2\sum_{\sigma_1=1}^{l_2}n_{\sigma_1};\\
&\sum_{\alpha_2=1}^{p_2}m_{\alpha_2}=2m_1,\ \sum_{\beta_2=1}^{q_2}n_{\beta_2}=2n_1+1;\\
&\sum_{\alpha_i=1}^{p}m_{\alpha_i}=2m_1,\ \sum_{\beta_i=1}^{q}n_{\beta_i}=2n_1+1,\ \text{for } i=1,2;
\end{aligned}\tag{1.99}$$

l_1, l_2–quadratic polynomials are without real roots, and the ρ_1^{th} and σ_1^{th}–quadratic polynomials are with powers m_{ρ_1} and n_{σ_1}, respectively.

(vi_1) For $a_{j_1j_1 0} > 0$ and $a_{j_2j_2 0} > 0$, there is a ([$(2m_1)$, $(2n_1+1)$]UP:UP,pF)-*positive hyperbolic-secant-to-hyperbolic flow as*

$$(a_{j_1j_1 1},a_{j_1j_2 1})=\underbrace{([(2m_1),(2n_1+1)]\text{UP:UP,pF})}_{[(2m_1),(2n_1+1)]\text{-hyperbolic-secant-to-hyperbolic flow }(+)}, \tag{1.100}$$

there are three following a ([$(2m_1)$, $(2n_1+1)$]UP:UP,pF)-*positive hyperbolic-secant-to-hyperbolic flow bifurcations.*

(vi_{1a}) The ([$(2m_1)$, $(2n_1+1)$]UP:UP,pF)-*positive hyperbolic-secant-to-hyperbolic flow appearing bifurcation is from a* (UP,pF)-*positive up-parabola flow to a hyperbolic and hyperbolic-secant flow network, i.e.,*

$$\underbrace{(a_{j_1j_2 1}, \dot{x}_{j_2})}_{\text{(UP, pF) - parabola flow}(+)} \rightleftharpoons \underbrace{([(2m_1), (2n_1+1)]\text{UP:UP,pF})}_{[(2m_1),(2n_1+1)]\text{-hyperbolic-secant-to-hyperbolic flow }(+)}$$
$$\rightleftharpoons \bigcup\nolimits_{\alpha=1}^{2m_1} \bigcup\nolimits_{\beta=1}^{2n_1+1} \underbrace{(a_{j_1j_1\alpha}, a_{j_1j_2\beta})}_{\text{(XX, pF)}}. \tag{1.101}$$

(vi_{1b}) The $([(2m_1), (2n_1+1)]$UP:UP,pF)*-positive hyperbolic-secant-to-hyperbolic flow appearing bifurcation is from a* $q_1 \times p_1$ *to* $q_2 \times p_2$ *singular hyperbolic and hyperbolic-secant flow networks, i.e.,*

$$\bigcup\nolimits_{\alpha_1=1}^{p_1} \bigcup\nolimits_{\beta_1=1}^{q_1} \underbrace{(a_{j_1j_1\alpha_1}, a_{j_1j_2\beta_1})}_{([m_{\alpha_1}, n_{\beta_1}]\text{XX, pF})} \rightleftharpoons \underbrace{([(2m_1), (2n_1+1)]\text{UP:UP,pF})}_{[(2m_1),(2n_1+1)]\text{-hyperbolic-secant-to-hyperbolic flow }(+)}$$
$$\rightleftharpoons \bigcup\nolimits_{\alpha_2=1}^{p_2} \bigcup\nolimits_{\beta_2=1}^{q_2} \underbrace{(a_{j_1j_1\alpha_2}, a_{j_1j_2\beta_2})}_{([m_{\alpha_2}, n_{\beta_2}]\text{XX, pF})}. \tag{1.102}$$

(vi_{1c}) The $([(2m_1), (2n_1+1)]$UP:UP,pF)*-positive hyperbolic-secant-to-hyperbolic flow switching bifurcation is for two* $q \times p$ *singular hyperbolic and hyperbolic-secant flow networks, i.e.,*

$$\bigcup\nolimits_{\alpha_1=1}^{p} \bigcup\nolimits_{\beta_1=1}^{q} \underbrace{(a_{j_1j_1\alpha_1}, a_{j_1j_2\beta_1})}_{([m_{\alpha_1}, n_{\beta_1}]\text{XX, pF})} \rightleftharpoons \underbrace{([(2m_1), (2n_1+1)]\text{UP:UP,pF})}_{[(2m_1),(2n_1+1)]\text{-hyperbolic-secant-to-hyperbolic flow }(+)}$$
$$\rightleftharpoons \bigcup\nolimits_{\alpha_2=1}^{p} \bigcup\nolimits_{\beta_2=1}^{q} \underbrace{(a_{j_1j_1\alpha_2}, a_{j_1j_2\beta_2})}_{([m_{\alpha_2}, n_{\beta_2}]\text{XX, pF})}. \tag{1.103}$$

(vi_2) For $a_{j_1j_1 0} < 0$ *and* $a_{j_2j_2 0} > 0$*, there is a* $([(2m_1), (2n_1+1)]$DP:DP,pF)*-positive hyperbolic-to-hyperbolic-secant flow as*

$$(a_{j_1j_1 1}, a_{j_1j_2 1}) = \underbrace{([(2m_1), (2n_1+1)]\text{DP:DP,pF})}_{[(2m_1),(2n_1+1)]\text{-hyperbolic-to-hyperbolic-secant flow }(+)}, \tag{1.104}$$

there are three following a $([(2m_1), (2n_1+1)]$DP:DP,pF)*-positive hyperbolic-to-hyperbolic-secant flow appearing and switching bifurcations.*

(vi_{2a}) The $([(2m_1), (2n_1+1)]$DP:DP,pF)*-positive hyperbolic-to-hyperbolic-secant flow appearing bifurcation is from a* (DP,pF)*-positive down-parabola flow to a hyperbolic and hyperbolic-secant flow network.*

$$\underbrace{(a_{j_1j_2 1}, \dot{x}_{j_2})}_{\text{(DP, pF)-down-parabola flow}(+)} \rightleftharpoons \underbrace{([(2m_1), (2n_1+1)]\text{DP:DP,pF})}_{[(2m_1),(2n_1+1)]\text{-hyperbolic-to-hyperbolic-secant flow }(+)}$$

$$\rightleftharpoons \bigcup\nolimits_{\alpha=1}^{2m_1} \bigcup\nolimits_{\beta=1}^{2n_1+1} \underbrace{(a_{j_1 j_1 \alpha}, a_{j_1 j_2 \beta})}_{\text{(XX, pF)}}. \tag{1.105}$$

(vi_{2b}) The ([(2m_1), (2n_1 + 1)]DP:DP,pF)*-positive hyperbolic-to-hyperbolic-secant flow appearing bifurcation is from a $q_1 \times p_1$ to $q_2 \times p_2$ singular hyperbolic and hyperbolic-secant flow networks*

$$\begin{aligned} \bigcup\nolimits_{\alpha_1=1}^{p_1} \bigcup\nolimits_{\beta_1=1}^{q_1} \underbrace{(a_{j_1 j_1 \alpha_1}, a_{j_1 j_2 \beta_1})}_{([m_{\alpha_1}, n_{\beta_1}]\text{XX, pF})} &\rightleftharpoons \underbrace{([(2m_1), (2n_1+1)]\text{DP:DP,pF})}_{[(2m_1),(2n_1+1)]\text{-hyperbolic-to-hyperbolic-secant flow }(+)} \\ &\rightleftharpoons \bigcup\nolimits_{\alpha_2=1}^{p_2} \bigcup\nolimits_{\beta_2=1}^{q_2} \underbrace{(a_{j_1 j_1 \alpha_2}, a_{j_1 j_2 \beta_2})}_{([m_{\alpha_2}, n_{\beta_2}]\text{XX, pF})}. \end{aligned} \tag{1.106}$$

(vi_{2c}) The ([(2m_1), (2n_1 + 1)]DP:DP,pF)*-positive hyperbolic-to-hyperbolic-secant flow switching bifurcation is for two $q \times p$ singular hyperbolic and hyperbolic-secant flow networks*

$$\begin{aligned} \bigcup\nolimits_{\alpha_1=1}^{p} \bigcup\nolimits_{\beta_1=1}^{q} \underbrace{(a_{j_1 j_1 \alpha_1}, a_{j_1 j_2 \beta_1})}_{([m_{\alpha_1}, n_{\beta_1}]\text{XX, pF})} &\rightleftharpoons \underbrace{([(2m_1), (2n_1+1)]\text{DP:DP,pF})}_{[(2m_1),(2n_1+1)]\text{-hyperbolic-to-hyperbolic-secant flow }(+)} \\ &\rightleftharpoons \bigcup\nolimits_{\alpha_2=1}^{p} \bigcup\nolimits_{\beta_2=1}^{q} \underbrace{(a_{j_1 j_1 \alpha_2}, a_{j_1 j_2 \beta_2})}_{([m_{\alpha_2}, n_{\beta_2}]\text{XX, pF})}. \end{aligned} \tag{1.107}$$

(vi_3) For $a_{j_1 j_1 0} > 0$ and $a_{j_2 j_2 0} < 0$, there is a ([(2m_1), (2n_1 + 1)]DP:DP,nF)*-negative hyperbolic-to-hyperbolic-secant flow as*

$$(a_{j_1 j_1 1}, a_{j_1 j_2 1}) = \underbrace{([(2m_1), (2n_1+1)]\text{DP:DP,nF})}_{[(2m_1),(2n_1+1)]\text{-hyperbolic-to-hyperbolic-secant flow }(-)}, \tag{1.108}$$

there are three following ([(2m_1), (2n_1 + 1)]DP:DP,nF)*-negative hyperbolic-to-hyperbolic-secant flow appearing and switching bifurcations.*

(vi_{3a}) The ([(2m_1), (2n_1 + 1)]DP:DP,nF)*-negative hyperbolic-to-hyperbolic-secant flow appearing bifurcation is from a* (DP,nF)*-negative down-parabola flow to a hyperbolic and hyperbolic-secant flow network.*

$$\begin{aligned} \underbrace{(a_{j_1 j_2 1}, \dot{x}_{j_2})}_{\text{(DP, nF)-down-parabola flow}} &\rightleftharpoons \underbrace{([(2m_1), (2n_1+1)]\text{DP:DP,nF})}_{[(2m_1),(2n_1+1)]\text{-hyperbolic-to-hyperbolic-secant flow }(-)} \\ &\rightleftharpoons \bigcup\nolimits_{\alpha=1}^{2m_1} \bigcup\nolimits_{\beta=1}^{2n_1+1} \underbrace{(a_{j_1 j_1 \alpha}, a_{j_1 j_2 \beta})}_{\text{(XX, nF)}}. \end{aligned} \tag{1.109}$$

(vi_{3b}) The ([(2m_1), (2n_1 + 1)]DP:DP,nF)*-negative hyperbolic-to-hyperbolic-secant flow appearing bifurcation is from a $q_1 \times p_1$ to $q_2 \times p_2$ singular hyperbolic and hyperbolic-secant flow networks*

$$\bigcup_{\alpha_1=1}^{p_1}\bigcup_{\beta_1=1}^{q_1}\underbrace{(a_{j_1j_1\alpha_1},a_{j_1j_2\beta_1})}_{([m_{\alpha_1},n_{\beta_1}]\mathrm{XX,nF})} \rightleftharpoons \underbrace{([(2m_1),(2n_1+1)]\mathrm{DP{:}DP,nF})}_{[(2m_1),(2n_1+1)]\text{-hyperbolic-to-hyperbolic-secant flow }(-)} \rightleftharpoons \bigcup_{\alpha_2=1}^{p_2}\bigcup_{\beta_2=1}^{q_2}\underbrace{(a_{j_1j_1\alpha_2},a_{j_1j_2\beta_2})}_{([m_{\alpha_2},n_{\beta_2}]\mathrm{XX,nF})}. \tag{1.110}$$

(vi_{3c}) The ([(2m_1), (2n_1 + 1)]DP:DP,nF)*-positive hyperbolic flow switching bifurcation is for two $q \times p$ singular hyperbolic and hyperbolic-secant flow networks*

$$\bigcup_{\alpha_1=1}^{p}\bigcup_{\beta_1=1}^{q}\underbrace{(a_{j_1j_1\alpha_1},a_{j_1j_2\beta_1})}_{([m_{\alpha_1},n_{\beta_1}]\mathrm{XX,nF})} \rightleftharpoons \underbrace{([(2m_1),(2n_1+1)]\mathrm{DP{:}DP,nF})}_{[(2m_1),(2n_1+1)]\text{-hyperbolic-to-hyperbolic-secant flow }(-)} \rightleftharpoons \bigcup_{\alpha_2=1}^{p}\bigcup_{\beta_2=1}^{q}\underbrace{(a_{j_1j_1\alpha_2},a_{j_1j_2\beta_2})}_{([m_{\alpha_2},n_{\beta_2}]\mathrm{XX,nF})}. \tag{1.111}$$

(vi_4) For $a_{j_1j_1 0} < 0$ and $a_{j_2j_2 0} < 0$, there is a ([(2m_1), (2n_1 + 1)]UP:UP,nF)*-negative hyperbolic-secant-to-hyperbolic flow as*

$$(a_{j_1j_1 1},a_{j_1j_2 1}) = \underbrace{([(2m_1),(2n_1+1)]\mathrm{UP{:}UP,nF})}_{[(2m_1),(2n_1+1)]\text{-hyperbolic-secant-to-hyperbolic flow }(-)}, \tag{1.112}$$

there are three following ([(2m_1), (2n_1 + 1)]DP:DP,nF)*-negative hyperbolic-to-hyperbolic-secant flow appearing and switching bifurcations.*

(vi_{4a}) The ([(2m_1), (2n_1 + 1)]DP:UP,nF)*-negative hyperbolic-secant-to-hyperbolic flow appearing bifurcation is from a* (UP,nF)*-negative up-parabola flow to a hyperbolic and hyperbolic-secant flow network, i.e.,*

$$\underbrace{(a_{j_1j_2 1},\dot{x}_{j_2})}_{(\mathrm{UP,\ nF})\text{-up-parabola flow}} \rightleftharpoons \underbrace{([(2m_1),(2n_1+1)]\mathrm{UP{:}UP,nF})}_{[(2m_1),(2n_1+1)]\text{-hyperbolic-secant-to-hyperbolic flow }(-)} \rightleftharpoons \bigcup_{\alpha=1}^{2m_1}\bigcup_{\beta=1}^{2n_1+1}\underbrace{(a_{j_1j_1\alpha},a_{j_1j_2\beta})}_{(\mathrm{XX,\ nF})}. \tag{1.113}$$

(vi_{4b}) The ([(2m_1), (2n_1 + 1)]DP:UP,nF)*-negative hyperbolic-secant-to-hyperbolic flow appearing bifurcation is from a $q_1 \times p_1$ to $q_2 \times p_2$ singular hyperbolic and hyperbolic-secant flow networks, i.e.,*

$$
\begin{aligned}
&\bigcup\nolimits_{\alpha_1=1}^{p_1}\bigcup\nolimits_{\beta_1=1}^{q_1}\underbrace{(a_{j_1j_1\alpha_1},a_{j_1j_2\beta_1})}_{([m_{\alpha_1},n_{\beta_1}]\text{XX, nF})}\rightleftharpoons\underbrace{([(2m_1),(2n_1+1)]\text{UP:UP,nF})}_{[(2m_1),(2n_1+1)]\text{-hyperbolic-secant-to-hyperbolic flow }(-)}\\
&\qquad\rightleftharpoons\bigcup\nolimits_{\alpha_2=1}^{p_2}\bigcup\nolimits_{\beta_2=1}^{q_2}\underbrace{(a_{j_1j_1\alpha_2},a_{j_1j_2\beta_2})}_{([m_{\alpha_2},n_{\beta_2}]\text{XX, nF})}.
\end{aligned}\tag{1.114}
$$

(vi_{4c}) The ([(2m_1), (2n_1 + 1)]DP:UP,nF)*-negative hyperbolic-secant-to-hyperbolic flow switching bifurcation is for two $q\times p$ singular hyperbolic and hyperbolic-secant flow networks, i.e.,*

$$
\begin{aligned}
&\bigcup\nolimits_{\alpha_1=1}^{p}\bigcup\nolimits_{\beta_1=1}^{q}\underbrace{(a_{j_1j_1\alpha_1},a_{j_1j_2\beta_1})}_{([m_{\alpha_1},n_{\beta_1}]\text{XX, nF})}\rightleftharpoons\underbrace{([(2m_1),(2n_1+1)]\text{UP:UP,nF})}_{[(2m_1),(2n_1+1)]\text{-hyperbolic-secant-to-hyperbolic flow }(-)}\\
&\qquad\rightleftharpoons\bigcup\nolimits_{\alpha_2=1}^{p}\bigcup\nolimits_{\beta_2=1}^{q}\underbrace{(a_{j_1j_1\alpha_2},a_{j_1j_2\beta_2})}_{([m_{\alpha_2},n_{\beta_2}]\text{XX, nF})}.
\end{aligned}\tag{1.115}
$$

(vii) For $m=2m_1+1$ and $n=2n_1$, the equilibrium of $(x_{j_1}^*,x_{j_2}^*)=(a_{j_1j_1 1},a_{j_1j_2 1})$ for the x_{j_1}-direction has the following bifurcation properties with

$$
\begin{aligned}
2m_1+1-\sum\nolimits_{\alpha_1=1}^{p_1}m_{\alpha_1}&=2\sum\nolimits_{\rho_1=1}^{l_1}m_{\rho_1},\;2n_1-\sum\nolimits_{\beta_1=1}^{q_1}n_{\beta_1}=2\sum\nolimits_{\sigma_1=1}^{l_2}n_{\sigma_1};\\
\sum\nolimits_{\alpha_2=1}^{p_2}m_{\alpha_2}&=2m_1+1,\;\sum\nolimits_{\beta_2=1}^{q_2}n_{\beta_2}=2n_1;\\
\sum\nolimits_{\alpha_i=1}^{p}m_{\alpha_i}&=2m_1+1,\;\sum\nolimits_{\beta_i=1}^{q}n_{\beta_i}=2n_1,\;\text{for } i=1,2;
\end{aligned}\tag{1.116}
$$

l_1,l_2-quadratic polynomials are without real roots, and the ρ_1^{th} and σ_1^{th}-quadratic polynomials are with powers m_{ρ_1} and n_{σ_1},respectively.

(vii_1) For $a_{j_1j_10}>0$ and $a_{j_2j_20}>0$, there is a ([(2m_1+1), (2n_1)]$_{\text{DI:II}}$SO,pF)*-positive decreasing-to-increasing inflection source flow as*

$$
(a_{j_1j_1 1},a_{j_1j_2 1})=\underbrace{([(2m_1+1),(2n_1)]_{\text{DI:II}}\text{SO,pF})}_{[(2m_1+1),(2n_1)]\text{-decreasing-to-increasing inflection source flow }(+)},\tag{1.117}
$$

there are three following ([(2m_1+1), (2n_1)]$_{\text{DI:II}}$SO,pF)*-positive decreasing-to-increasing inflection source flow appearing and switching bifurcations.*

(vii_{1a}) The ([(2m_1+1), (2n_1)]$_{\text{DI:II}}$SO,pF)*-positive decreasing-to-increasing inflection source flow appearing bifurcation is from an* (SO,pF)*-positive source flow to a hyperbolic and hyperbolic-secant flow network, i.e.,*

$$\underbrace{(a_{j_1j_11}, \dot{x}_{j_2})}_{\text{(SO, pF)-source flow}} \rightleftharpoons \underbrace{([(2m_1+1), (2n_1)]_{\text{DI:II}}\text{SO,pF})}_{[(2m_1+1),(2n_1)]\text{-decreasing-to-increasing inflection source flow }(+)}$$
$$\rightleftharpoons \bigcup\nolimits_{\alpha=1}^{2m_1+1} \bigcup\nolimits_{\beta=1}^{2n_1} \underbrace{(a_{j_1j_1\alpha}, a_{j_1j_2\beta})}_{\text{(XX, pF)}}. \tag{1.118}$$

(vii$_{1b}$) The ([(2m_1+1), (2n_1)]$_{\text{DI:II}}$SO,pF)*-positive decreasing-to-increasing inflection source flow appearing bifurcation is from a $q_1 \times p_1$ to $q_2 \times p_2$ singular hyperbolic and hyperbolic-secant flow networks, i.e.,*

$$\bigcup\nolimits_{\alpha_1=1}^{p_1} \bigcup\nolimits_{\beta_1=1}^{q_1} \underbrace{(a_{j_1j_1\alpha_1}, a_{j_1j_2\beta_1})}_{([m_{\alpha_1}, n_{\beta_1}]\text{XX, pF})} \rightleftharpoons \underbrace{([(2m_1+1), (2n_1)]_{\text{DI:II}}\text{SO,pF})}_{[(2m_1+1),(2n_1)]\text{-decreasing-to-increasing inflection source flow }(+)}$$
$$\rightleftharpoons \bigcup\nolimits_{\alpha_2=1}^{p_2} \bigcup\nolimits_{\beta_2=1}^{q_2} \underbrace{(a_{j_1j_1\alpha_2}, a_{j_1j_2\beta_2})}_{([m_{\alpha_2}, n_{\beta_2}]\text{XX, pF})}. \tag{1.119}$$

(vii$_{1c}$) The ([(2m_1+1), (2n_1)]$_{\text{DI:II}}$SO,pF)*-positive decreasing-to-increasing inflection source flow switching bifurcation is for two $q \times p$ singular hyperbolic and hyperbolic-secant flow networks, i.e.,*

$$\bigcup\nolimits_{\alpha_1=1}^{p} \bigcup\nolimits_{\beta_1=1}^{q} \underbrace{(a_{j_1j_1\alpha_1}, a_{j_1j_2\beta_1})}_{([m_{\alpha_1}, n_{\beta_1}]\text{XX,xF})} \rightleftharpoons \underbrace{([(2m_1+1), (2n_1)]_{\text{DI:II}}\text{SO,pF})}_{[(2m_1+1),(2n_1)]\text{-decreasing-to-increasing inflection source flow }(+)}$$
$$\rightleftharpoons \bigcup\nolimits_{\alpha_2=1}^{p} \bigcup\nolimits_{\beta_2=1}^{q} \underbrace{(a_{j_1j_1\alpha_2}, a_{j_1j_2\beta_2})}_{([m_{\alpha_2}, n_{\beta_2}]\text{XX, pF})}. \tag{1.120}$$

(vii$_2$) For $a_{j_1j_10} < 0$ and $a_{j_2j_20} > 0$, there is a ([(2m_1+1), (2n_1)]$_{\text{II:DI}}$SI,pF)*-positive increasing-to-decreasing inflection sink flow as*

$$(a_{j_1j_11}, a_{j_1j_21}) = \underbrace{([(2m_1+1), (2n_1)]_{\text{II:DI}}\text{SI,pF})}_{[(2m_1+1),(2n_1)]\text{-increasing-to-decreasing inflection sink flow }(+)}, \tag{1.121}$$

there are three following ([(2m_1+1), (2n_1)]$_{\text{II:DI}}$SI,pF)*-positive increasing-to-decreasing inflection sink flow appearing and switching bifurcations.*

(vii$_{2a}$) The ([(2m_1+1), (2n_1)]$_{\text{II:DI}}$SI,pF)*-positive increasing-to-decreasing inflection sink flow appearing bifurcation is from an* (SI,pF)*-positive hyperbolic-secant flow to a hyperbolic and hyperbolic-secant flow network.*

$$\underbrace{(a_{j_1j_1 1}, \dot{x}_{j_2})}_{\text{(SI, pF)-sink flow}} \rightleftharpoons \underbrace{([(2m_1+1), (2n_1)]_{\text{II:DI}}\text{SI,pF})}_{[(2m_1+1),(2n_1)]\text{-increasing-to-decreasing inflection sink flow }(+)} \rightleftharpoons \bigcup_{\alpha=1}^{2m_1+1}\bigcup_{\beta=1}^{2n_1}\underbrace{(a_{j_1j_1\alpha}, a_{j_1j_2\beta})}_{\text{(XX, pF)}}. \tag{1.122}$$

(vii$_{2b}$) The $([(2m_1+1), (2n_1)]_{\text{II:DI}}\text{SI,pF})$*-positive increasing-to-decreasing inflection sink flow appearing bifurcation is from a* $q_1 \times p_1$ *to* $q_2 \times p_2$ *singular hyperbolic and hyperbolic-secant flow networks*

$$\bigcup_{\alpha_1=1}^{p_1}\bigcup_{\beta_1=1}^{q_1}\underbrace{(a_{j_1j_1\alpha_1}, a_{j_1j_2\beta_1})}_{([m_{\alpha_1},n_{\beta_1}]\text{XX, pF})} \rightleftharpoons \underbrace{([(2m_1+1), (2n_1)]_{\text{II:DI}}\text{SI,pF})}_{[(2m_1+1),(2n_1)]\text{-increasing-to-decreasing inflection sink flow }(+)} \rightleftharpoons \bigcup_{\alpha_2=1}^{p_2}\bigcup_{\beta_2=1}^{q_2}\underbrace{(a_{j_1j_1\alpha_2}, a_{j_1j_2\beta_2})}_{([m_{\alpha_2},n_{\beta_2}]\text{XX, pF})}. \tag{1.123}$$

(vii$_{2c}$) The $([(2m_1+1), (2n_1)]_{\text{II:DI}}\text{SI,pF})$*-positive increasing-to-decreasing inflection sink flow switching bifurcation is for two* $q \times p$ *singular hyperbolic and hyperbolic-secant flow networks*

$$\bigcup_{\alpha_1=1}^{p}\bigcup_{\beta_1=1}^{q}\underbrace{(a_{j_1j_1\alpha_1}, a_{j_1j_2\beta_1})}_{([m_{\alpha_1},n_{\beta_1}]\text{XX, pF})} \rightleftharpoons \underbrace{([(2m_1+1), (2n_1)]_{\text{II:DI}}\text{SI,pF})}_{[(2m_1+1),(2n_1)]\text{-increasing-to-decreasing inflection sink flow }(+)} \rightleftharpoons \bigcup_{\alpha_2=1}^{p}\bigcup_{\beta_2=1}^{q}\underbrace{(a_{j_1j_1\alpha_2}, a_{j_1j_2\beta_2})}_{([m_{\alpha_2},n_{\beta_2}]\text{XX, pF})}. \tag{1.124}$$

(vii$_3$) For $a_{j_1j_10} > 0$ *and* $a_{j_2j_20} < 0$*, there is a* $([(2m_1+1), (2n_1)]_{\text{II:DI}}\text{SO,nF})$*-negative increasing-to-decreasing inflection source flow as*

$$(a_{j_1j_1 1}, a_{j_1j_2 1}) = \underbrace{([(2m_1+1), (2n_1)]_{\text{II:DI}}\text{SO,nF})}_{[(2m_1+1),(2n_1)]\text{-increasing-to-decreasing inflection source flow }(-)}, \tag{1.125}$$

there are three following $([(2m_1+1), (2n_1)]_{\text{II:DI}}\text{SO,nF})$*-negative increasing-to-decreasing inflection source flow appearing and switching bifurcations.*

(vii$_{3a}$) The $([(2m_1+1), (2n_1)]_{\text{II:DI}}\text{SO,nF})$*-negative increasing-to-decreasing inflection source flow appearing bifurcation is from an* (SO,nF)*-negative source flow to a hyperbolic and hyperbolic-secant flow network.*

$$\underbrace{(a_{j_1j_1 1}, \dot{x}_{j_2})}_{\text{(SO, nF)-source flow}} \rightleftharpoons \underbrace{([(2m_1+1), (2n_1)]_{\text{II:DI}}\text{SO,nF})}_{[(2m_1+1),(2n_1)]\text{-increasing-to-decreasing inflection source flow }(-)}$$
$$\rightleftharpoons \bigcup\nolimits_{\alpha=1}^{2m_1+1} \bigcup\nolimits_{\beta=1}^{2n_1} \underbrace{(a_{j_1j_1\alpha}, a_{j_1j_2\beta})}_{\text{(XX, nF)}}. \tag{1.126}$$

(vii$_{3b}$) The ([(2m_1+1), (2n_1)]$_{\text{II:DI}}$SO,nF)*-negative increasing-to-decreasing inflection source flow appearing bifurcation is from a $q_1 \times p_1$ to $q_2 \times p_2$ singular hyperbolic and hyperbolic-secant flow networks*

$$\bigcup\nolimits_{\alpha_1=1}^{p_1} \bigcup\nolimits_{\beta_1=1}^{q_1} \underbrace{(a_{j_1j_1\alpha_1}, a_{j_1j_2\beta_1})}_{([m_{\alpha_1},n_{\beta_1}]\text{XX, nF})} \rightleftharpoons \underbrace{([(2m_1+1), (2n_1)]_{\text{II:DI}}\text{SO,nF})}_{[(2m_1+1),(2n_1)]\text{-increasing-to-decreasing inflection source flow }(-)}$$
$$\rightleftharpoons \bigcup\nolimits_{\alpha_2=1}^{p_2} \bigcup\nolimits_{\beta_2=1}^{q_2} \underbrace{(a_{j_1j_1\alpha_2}, a_{j_1j_2\beta_2})}_{([m_{\alpha_2},n_{\beta_2}]\text{XX, nF})}. \tag{1.127}$$

(vii$_{3c}$) The ([(2m_1+1), (2n_1)]$_{\text{II:DI}}$SO,nF)*-negative increasing-to-decreasing inflection source flow switching bifurcation is for two $q \times p$ singular hyperbolic and hyperbolic-secant flow networks*

$$\bigcup\nolimits_{\alpha_1=1}^{p} \bigcup\nolimits_{\beta_1=1}^{q} \underbrace{(a_{j_1j_1\alpha_1}, a_{j_1j_2\beta_1})}_{([m_{\alpha_1},n_{\beta_1}]\text{XX, nF})} \rightleftharpoons \underbrace{([(2m_1+1), (2n_1)]_{\text{II:DI}}\text{SO,nF})}_{[(2m_1+1),(2n_1)]\text{-increasing-to-decreasing inflection source flow }(-)}$$
$$\rightleftharpoons \bigcup\nolimits_{\alpha_2=1}^{p} \bigcup\nolimits_{\beta_2=1}^{q} \underbrace{(a_{j_1j_1\alpha_2}, a_{j_1j_2\beta_2})}_{([m_{\alpha_2},n_{\beta_2}]\text{XX, nF})}. \tag{1.128}$$

(vii$_4$) For $a_{j_1j_1 0} < 0$ and $a_{j_2j_2 0} < 0$, there is a ([(2m_1+1), (2n_1)]$_{\text{DI:II}}$SI,nF)*-negative decreasing-to-increasing inflection sink flow as*

$$(a_{j_1j_1 1}, a_{j_1j_2 1}) = \underbrace{([(2m_1+1), (2n_1)]_{\text{DI:II}}\text{SI,nF})}_{[(2m_1+1),(2n_1)]\text{-decreasing-to-increasing inflection sink flow }(-)}, \tag{1.129}$$

there are three following ([(2m_1+1), (2n_1)]$_{\text{DI:II}}$SI,nF)*-negative decreasing-to-increasing inflection sink flow appearing and switching bifurcations.*

(vii$_{4a}$) The ([(2m_1+1), (2n_1)]$_{\text{DI:II}}$SI,nF)*-negative decreasing-to-increasing inflection sink flow appearing bifurcation is from an* (SI,nF)*-negative sink flow to a hyperbolic and hyperbolic-secant flow network, i.e.,*

$$\underbrace{(a_{j_1j_11}, \dot{x}_{j_2})}_{\text{(SI, nF)-sink flow}} \rightleftharpoons \underbrace{([(2m_1+1), (2n_1)]_{\text{DI:II}}\text{SI,nF})}_{[(2m_1+1),(2n_1)]\text{-decreasing-to-increasing inflection sink flow }(-)}$$
$$\rightleftharpoons \bigcup\nolimits_{\alpha=1}^{2m_1+1} \bigcup\nolimits_{\beta=1}^{2n_1} \underbrace{(a_{j_1j_1\alpha}, a_{j_1j_2\beta})}_{\text{(XX, nF)}}. \tag{1.130}$$

(vii$_{4b}$) The $([(2m_1+1), (2n_1)]_{\text{DI:II}}\text{SI,nF})$*-negative decreasing-to-increasing inflection sink flow appearing bifurcation is from a $q_1 \times p_1$ to $q_2 \times p_2$ singular hyperbolic and hyperbolic-secant flow networks, i.e.,*

$$\bigcup\nolimits_{\alpha_1=1}^{p_1} \bigcup\nolimits_{\beta_1=1}^{q_1} \underbrace{(a_{j_1j_1\alpha_1}, a_{j_1j_2\beta_1})}_{([m_{\alpha_1}, n_{\beta_1}]\text{XX, nF})} \rightleftharpoons \underbrace{([(2m_1+1), (2n_1)]_{\text{DI:II}}\text{SI,nF})}_{[(2m_1+1),(2n_1)]\text{-decreasing-to-increasing inflection sink flow }(-)}$$
$$\rightleftharpoons \bigcup\nolimits_{\alpha_2=1}^{p_2} \bigcup\nolimits_{\beta_2=1}^{q_2} \underbrace{(a_{j_1j_1\alpha_2}, a_{j_1j_2\beta_2})}_{([m_{\alpha_2}, n_{\beta_2}]\text{XX, nF})}. \tag{1.131}$$

(vii$_{4c}$) The $([(2m_1+1), (2n_1)]_{\text{DI:II}}\text{SI,nF})$*-negative decreasing-to-increasing inflection sink flow switching bifurcation is for two $q \times p$ singular hyperbolic and hyperbolic-secant flow networks, i.e.,*

$$\bigcup\nolimits_{\alpha_1=1}^{p} \bigcup\nolimits_{\beta_1=1}^{q} \underbrace{(a_{j_1j_1\alpha_1}, a_{j_1j_2\beta_1})}_{([m_{\alpha_1}, n_{\beta_1}]\text{XX, nF})} \rightleftharpoons \underbrace{([(2m_1+1), (2n_1)]_{\text{DI:II}}\text{SI,nF})}_{[(2m_1+1),(2n_1)]\text{-decreasing-to-increasing inflection sink flow }(-)}$$
$$\rightleftharpoons \bigcup\nolimits_{\alpha_2=1}^{p} \bigcup\nolimits_{\beta_2=1}^{q} \underbrace{(a_{j_1j_1\alpha_2}, a_{j_1j_2\beta_2})}_{([m_{\alpha_2}, n_{\beta_2}]\text{XX, nF})}. \tag{1.132}$$

(viii) For $m = 2m_1$ and $n = 2n_1$, the equilibrium of $(x^*_{j_1}, x^*_{j_2}) = (a_{j_1j_11}, a_{j_1j_21})$ for the x_{j_1}-direction has the following bifurcation properties with

$$\begin{aligned}
&2m_1 - \sum\nolimits_{\alpha_1=1}^{p_1} m_{\alpha_1} = 2\sum\nolimits_{\rho_1=1}^{l_1} m_{\rho_1}, 2n_1 - \sum\nolimits_{\beta_1=1}^{q_1} n_{\beta_1} = 2\sum\nolimits_{\sigma_1=1}^{l_2} n_{\sigma_1};\\
&\sum\nolimits_{\alpha_2=1}^{p_2} m_{\alpha_2} = 2m_1, \sum\nolimits_{\beta_2=1}^{q_2} n_{\beta_2} = 2n_1;\\
&\sum\nolimits_{\alpha_i=1}^{p} m_{\alpha_i} = 2m_1, \sum\nolimits_{\beta_i=1}^{q} n_{\beta_i} = 2n_1, \text{ for } i = 1, 2;
\end{aligned} \tag{1.133}$$

l_1, l_2-quadratic polynomials are without real roots, and the ρ_1^{th} and σ_1^{th}-quadratic polynomials are with powers m_{ρ_1} and n_{σ_1}, respectively.

(viii$_1$) For $a_{j_1j_10} > 0$ and $a_{j_2j_20} > 0$, there is a $([(2m_1), (2n_1)]_{\text{II:II}}\text{US,pF})$*-positive increasing-inflection upper-saddle flow as*

$$(a_{j_1j_11}, a_{j_1j_21}) = \underbrace{([(2m_1), (2n_1)]_{\text{II:II}}\text{US,pF})}_{[(2m_1),(2n_1)]\text{-increasing-inflection upper-saddle flow }(+)}, \tag{1.134}$$

there are three following $([(2m_1), (2n_1)]_{\text{II:II}}\text{US,pF})$*-positive increasing-inflection upper-saddle flow appearing and switching bifurcations.*

($viii_{1a}$) The $([(2m_1), (2n_1)]_{\text{II:II}}\text{US,pF})$*-positive increasing-inflection upper-saddle flow appearing bifurcation is from a* (pF, pF)*-flow to a hyperbolic and hyperbolic-secant flow network, i.e.,*

$$\begin{aligned}\underbrace{(\dot{x}_{j_1}, \dot{x}_{j_2})}_{\text{(pF, pF)-flow}} &\rightleftharpoons \underbrace{([(2m_1), (2n_1)]_{\text{DI:II}}\text{US,pF})}_{[(2m_1),(2n_1)]\text{-increasing-inflection upper-saddle flow }(+)} \\ &\rightleftharpoons \bigcup\nolimits_{\alpha=1}^{2m_1} \bigcup\nolimits_{\beta=1}^{2n_1} \underbrace{(a_{j_1j_1\alpha}, a_{j_1j_2\beta})}_{\text{XX}}.\end{aligned} \tag{1.135}$$

($viii_{1b}$) The $([(2m_1), (2n_1)]_{\text{II:II}}\text{US,pF})$*-positive increasing-inflection upper-saddle flow appearing bifurcation is from a* $q_1 \times p_1$ *to* $q_2 \times p_2$ *singular hyperbolic and hyperbolic-secant flow networks, i.e.,*

$$\begin{aligned}\bigcup\nolimits_{\alpha_1=1}^{p_1} \bigcup\nolimits_{\beta_1=1}^{q_1} \underbrace{(a_{j_1j_1\alpha_1}, a_{j_1j_2\beta_1})}_{([m_{\alpha_1}, n_{\beta_1}]\text{XX, pF})} &\rightleftharpoons \underbrace{([(2m_1), (2n_1)]_{\text{II:II}}\text{US,pF})}_{[(2m_1),(2n_1)]\text{-increasing-inflection upper-saddle flow }(+)} \\ &\rightleftharpoons \bigcup\nolimits_{\alpha_2=1}^{p_2} \bigcup\nolimits_{\beta_2=1}^{q_2} \underbrace{(a_{j_1j_1\alpha_2}, a_{j_1j_2\beta_2})}_{([m_{\alpha_2}, n_{\beta_2}]\text{XX, pF})}.\end{aligned} \tag{1.136}$$

($viii_{1c}$) The $([(2m_1), (2n_1)]_{\text{II:II}}\text{US,pF})$*-positive increasing-inflection upper-saddle flow switching bifurcation is for two* $q \times p$ *singular hyperbolic and hyperbolic-secant flow networks, i.e.,*

$$\begin{aligned}\bigcup\nolimits_{\alpha_1=1}^{p} \bigcup\nolimits_{\beta_1=1}^{q} \underbrace{(a_{j_1j_1\alpha_1}, a_{j_1j_2\beta_1})}_{([m_{\alpha_1}, n_{\beta_1}]\text{XX, pF})} &\rightleftharpoons \underbrace{([(2m_1), (2n_1)]_{\text{II:II}}\text{US,pF})}_{[(2m_1),(2n_1)]\text{-increasing-inflection upper-saddle flow }(+)} \\ &\rightleftharpoons \bigcup\nolimits_{\alpha_2=1}^{p} \bigcup\nolimits_{\beta_2=1}^{q} \underbrace{(a_{j_1j_1\alpha_2}, a_{j_1j_2\beta_2})}_{([m_{\alpha_2}, n_{\beta_2}]\text{XX, pF})}.\end{aligned} \tag{1.137}$$

($viii_2$) For $a_{j_1j_10} < 0$ *and* $a_{j_2j_20} > 0$*, there is a* $([(2m_1), (2n_1)]_{\text{DI:DI}}\text{LS,pF})$*-positive decreasing-inflection lower-saddle flow as*

$$(a_{j_1j_11}, a_{j_1j_21}) = \underbrace{([(2m_1), (2n_1)]_{\text{DI:DI}}\text{LS,pF})}_{[(2m_1),(2n_1)]\text{-decreasing-inflection lower-saddle flow }(+)}, \tag{1.138}$$

there are three following ([(2m_1), (2n_1)]$_{\text{DI:DI}}$LS,pF)*-positive decreasing-inflection lower-saddle flow appearing and switching bifurcations.*

(viii$_{2a}$) The ([(2m_1), (2n_1)]$_{\text{DI:DI}}$LS,pF)*-positive decreasing-inflection lower-saddle flow appearing bifurcation is from a* (nF,pF)*-flow to a hyperbolic and hyperbolic-secant flow network.*

$$\underbrace{(\dot{x}_{j_1}, \dot{x}_{j_2})}_{\text{(nF, pF) - flow}} \rightleftharpoons \underbrace{([(2m_1), (2n_1)]_{\text{DI:DI}}\text{LS,pF})}_{[(2m_1),(2n_1)]\text{-decreasing-inflection lower-saddle flow }(+)} \rightleftharpoons \bigcup_{\alpha=1}^{2m_1} \bigcup_{\beta=1}^{2n_1} \underbrace{(a_{j_1 j_1 \alpha}, a_{j_1 j_2 \beta})}_{\text{(XX, pF)}}. \tag{1.139}$$

(viii$_{2b}$) The ([(2m_1), (2n_1)]$_{\text{DI:DI}}$LS,pF)*-positive decreasing-inflection lower-saddle flow appearing bifurcation is from a* $q_1 \times p_1$ *to* $q_2 \times p_2$ *singular hyperbolic and hyperbolic-secant flow networks*

$$\bigcup_{\alpha_1=1}^{p_1} \bigcup_{\beta_1=1}^{q_1} \underbrace{(a_{j_1 j_1 \alpha_1}, a_{j_1 j_2 \beta_1})}_{([m_{\alpha_1}, n_{\beta_1}]\text{XX, pF})} \rightleftharpoons \underbrace{([(2m_1), (2n_1)]_{\text{DI:DI}}\text{LS,pF})}_{[(2m_1),(2n_1)]\text{-decreasing-inflection lower-saddle flow }(+)} \rightleftharpoons \bigcup_{\alpha_2=1}^{p_2} \bigcup_{\beta_2=1}^{q_2} \underbrace{(a_{j_1 j_1 \alpha_2}, a_{j_1 j_2 \beta_2})}_{([m_{\alpha_2}, n_{\beta_2}]\text{XX, pF})}. \tag{1.140}$$

(viii$_{2c}$) The ([(2m_1), (2n_1)]$_{\text{DI:DI}}$LS,pF)*-positive decreasing-inflection lower-saddle flow switching bifurcation is for two* $q \times p$ *singular hyperbolic and hyperbolic-secant flow networks*

$$\bigcup_{\alpha_1=1}^{p} \bigcup_{\beta_1=1}^{q} \underbrace{(a_{j_1 j_1 \alpha_1}, a_{j_1 j_2 \beta_1})}_{([m_{\alpha_1}, n_{\beta_1}]\text{XX, pF})} \rightleftharpoons \underbrace{([(2m_1), (2n_1)]_{\text{DI:DI}}\text{LS,pF})}_{[(2m_1),(2n_1)]\text{-decreasing-inflection lower-saddle flow }(+)} \rightleftharpoons \bigcup_{\alpha_2=1}^{p} \bigcup_{\beta_2=1}^{q} \underbrace{(a_{j_1 j_1 \alpha_2}, a_{j_1 j_2 \beta_2})}_{([m_{\alpha_2}, n_{\beta_2}]\text{XX, pF})}. \tag{1.141}$$

(viii$_3$) For $a_{j_1 j_1 0} > 0$ *and* $a_{j_2 j_2 0} < 0$*, there is a* ([(2m_1), (2n_1)]$_{\text{DI:DI}}$US,nF)*-negative decreasing-inflection upper-saddle flow as*

$$(a_{j_1 j_1 1}, a_{j_1 j_2 1}) = \underbrace{([(2m_1), (2n_1)]_{\text{DI:DI}}\text{US,nF})}_{[(2m_1),(2n_1)]\text{-decreasing-inflection upper-saddle flow }(-)}, \tag{1.142}$$

there are three following ([(2m_1), (2n_1)]$_{\text{DI:DI}}$US,nF)*-negative decreasing-inflection upper-saddle flow appearing and switching bifurcations.*

(viii$_{3a}$) The $([(2m_1),(2n_1)]_{\text{DI:DI}}\text{US,nF})$*-negative decreasing-inflection upper-saddle flow appearing bifurcation is from a* (pF, nF)*-flow to a hyperbolic and hyperbolic-secant flow network.*

$$\underbrace{(\dot{x}_{j_1},\dot{x}_{j_2})}_{\text{(pF, nF)-flow}} \rightleftharpoons \underbrace{([(2m_1),(2n_1)]_{\text{DI:DI}}\text{US,nF})}_{[(2m_1),(2n_1)]\text{-decreasing-inflection upper saddle flow }(-)} \rightleftharpoons \bigcup\nolimits_{\alpha=1}^{2m_1}\bigcup\nolimits_{\beta=1}^{2n_1}\underbrace{(a_{j_1j_1\alpha},a_{j_1j_2\beta})}_{\text{(XX, nF)}}. \tag{1.143}$$

(viii$_{3b}$) The $([(2m_1),(2n_1)]_{\text{DI:DI}}\text{US,nF})$*-negative decreasing-inflection upper-saddle flow appearing bifurcation is from a* $q_1\times p_1$ *to* $q_2\times p_2$ *singular hyperbolic and hyperbolic-secant flow networks*

$$\bigcup\nolimits_{\alpha_1=1}^{p_1}\bigcup\nolimits_{\beta_1=1}^{q_1}\underbrace{(a_{j_1j_1\alpha_1},a_{j_1j_2\beta_1})}_{([m_{\alpha_1},n_{\beta_1}]\text{XX, nF})} \rightleftharpoons \underbrace{([(2m_1),(2n_1)]_{\text{DI:DI}}\text{US,nF})}_{[(2m_1),(2n_1)]\text{-decreasing-inflection upper-saddle flow }(-)} \rightleftharpoons \bigcup\nolimits_{\alpha_2=1}^{p_2}\bigcup\nolimits_{\beta_2=1}^{q_2}\underbrace{(a_{j_1j_1\alpha_2},a_{j_1j_2\beta_2})}_{([m_{\alpha_2},n_{\beta_2}]\text{XX, nF})}. \tag{1.144}$$

(viii$_{3c}$) The $([(2m_1),(2n_1)]_{\text{DI:DI}}\text{US,nF})$*-negative decreasing-inflection upper-saddle flow switching bifurcation is for two* $q\times p$ *singular hyperbolic and hyperbolic-secant flow networks*

$$\bigcup\nolimits_{\alpha_1=1}^{p}\bigcup\nolimits_{\beta_1=1}^{q}\underbrace{(a_{j_1j_1\alpha_1},a_{j_1j_2\beta_1})}_{([m_{\alpha_1},n_{\beta_1}]\text{XX, nF})} \rightleftharpoons \underbrace{([(2m_1),(2n_1)]_{\text{DI:DI}}\text{US,nF})}_{[(2m_1),(2n_1)]\text{-decreasing-inflection upper-saddle flow }(-)} \rightleftharpoons \bigcup\nolimits_{\alpha_2=1}^{p}\bigcup\nolimits_{\beta_2=1}^{q}\underbrace{(a_{j_1j_1\alpha_2},a_{j_1j_2\beta_2})}_{([m_{\alpha_2},n_{\beta_2}]\text{XX, nF})}. \tag{1.145}$$

(viii$_4$) For $a_{j_1j_10}<0$ *and* $a_{j_2j_20}<0$*, there is a* $([(2m_1),(2n_1)]_{\text{II:II}}\text{LS,nF})$*-negative increasing-inflection lower-saddle flow as*

$$(a_{j_1j_11},a_{j_1j_21}) = \underbrace{([(2m_1),(2n_1)]_{\text{II:II}}\text{LS,nF})}_{[(2m_1),(2n_1)]\text{-increasing-inflection lower-saddle flow }(-)}, \tag{1.146}$$

there are three following $([(2m_1),(2n_1)]_{\text{II:II}}\text{LS,nF})$*-negative increasing-inflection lower-saddle flow appearing and switching bifurcations.*

(viii$_{4a}$) The $([(2m_1),(2n_1)]_{\text{II:II}}\text{LS,nF})$*-negative increasing-inflection lower-saddle flow appearing bifurcation is from an* (nF,nF)*-negative sink flow to a hyperbolic and hyperbolic-secant flow network, i.e.,*

$$\underbrace{(\dot{x}_{j_1}, \dot{x}_{j_2})}_{\text{(nF, nF)-flow}} \rightleftharpoons \underbrace{([(2m_1), (2n_1)]_{\text{II:II}}\text{LS,nF})}_{[(2m_1),(2n_1)]\text{-increasing-inflection lower-saddle flow }(-)}$$
$$\rightleftharpoons \bigcup\nolimits_{\alpha=1}^{2m_1} \bigcup\nolimits_{\beta=1}^{2n_1} \underbrace{(a_{j_1 j_1 \alpha}, a_{j_1 j_2 \beta})}_{\text{(XX, nF)}}. \tag{1.147}$$

(viii$_{4b}$) The $([(2m_1), (2n_1)]_{\text{II:II}}\text{LS,nF})$*-negative increasing-inflection lower-saddle flow appearing bifurcation is from a* $q_1 \times p_1$ *to* $q_2 \times p_2$ *singular hyperbolic and hyperbolic-secant flow networks, i.e.,*

$$\bigcup\nolimits_{\alpha_1=1}^{p_1} \bigcup\nolimits_{\beta_1=1}^{q_1} \underbrace{(a_{j_1 j_1 \alpha_1}, a_{j_1 j_2 \beta_1})}_{([m_{\alpha_1}, n_{\beta_1}]\text{XX, nF})} \rightleftharpoons \underbrace{([(2m_1), (2n_1)]_{\text{II:II}}\text{LS,nF})}_{[(2m_1),(2n_1)]\text{-increasing-inflection lower-saddle flow }(-)}$$
$$\rightleftharpoons \bigcup\nolimits_{\alpha_2=1}^{p_2} \bigcup\nolimits_{\beta_2=1}^{q_2} \underbrace{(a_{j_1 j_1 \alpha_2}, a_{j_1 j_2 \beta_2})}_{([m_{\alpha_2}, n_{\beta_2}]\text{XX, nF})}. \tag{1.148}$$

(viii$_{4c}$) The $([(2m_1), (2n_1)]_{\text{II:II}}\text{LS,nF})$*-negative increasing-inflection lower-saddle flow switching bifurcation is for two* $q \times p$ *singular hyperbolic and hyperbolic-secant flow networks, i.e.,*

$$\bigcup\nolimits_{\alpha_1=1}^{p} \bigcup\nolimits_{\beta_1=1}^{q} \underbrace{(a_{j_1 j_1 \alpha_1}, a_{j_1 j_2 \beta_1})}_{([m_{\alpha_1}, n_{\beta_1}]\text{XX, nF})} \rightleftharpoons \underbrace{([(2m_1), (2n_1)]_{\text{II:II}}\text{LS,nF})}_{[(2m_1),(2n_1)]\text{-decreasing-inflection lower-saddle flow }(-)}$$
$$\rightleftharpoons \bigcup\nolimits_{\alpha_2=1}^{p} \bigcup\nolimits_{\beta_2=1}^{q} \underbrace{(a_{j_1 j_1 \alpha_2}, a_{j_1 j_2 \beta_2})}_{([m_{\alpha_2}, n_{\beta_2}]\text{XX, nF})}. \tag{1.149}$$

Chapter 2
Proof of Theorem 1.1

In this chapter, the proof of the Theorem 1.1 is completed in detail. The sink and source switch and lower-saddle and upper-saddle switch are proved first. The results of singular hyperbolic flows, inflection-sink, source, and saddle flows, hyperbolic to hyperbolic-secant flows are proved, and the properties of networks of hyperbolic flows, inflection-sink, source, and saddle flows, hyperbolic-to-hyperbolic-secant flows are determined. The networks of hyperbolic and hyperbolic-secant flows are proved. The appearing and switching bifurcations conditions among the networks of aforementioned networks are completely obtained.

2.1 Sink and Source Switch, Lower and Supper-Saddle Switch

Consider a two-dimensional, nonlinear dynamical system as

$$\begin{aligned}\dot{x}_{j_2} &= a_{j_2j_20},\\ \dot{x}_{j_1} &= a_{j_1j_10}(x_{j_1} - a_{j_1j_11})^m(x_{j_2} - a_{j_1j_21})^n.\end{aligned}$$

If $a_{j_2j_20} = 0$, the 2-dimensional dynamical system is a 1-dimesional system in the x_{j_1}-direction. The foregoing equation becomes

$$\begin{aligned}dx_{j_2} &= 0 \cdot dt,\\ \frac{dx_{j_1}}{(x_{j_1} - a_{j_1j_11})^m} &= a_{j_1j_10}(x_{j_2} - a_{j_1j_21})^n dt.\end{aligned}$$

With an initial condition (x_{j_10}, x_{j_20}) art time $t = t_0$, integration of the foregoing equation gives the corresponding solution as

A. C. J. Luo, *Two-dimensional Constant and Product Polynomial Systems*,
https://doi.org/10.1007/978-981-96-5515-1_2

$$
\begin{aligned}
& x_{j_2} = x_{j_2 0}; \\
& -\frac{1}{m-1} a_{j_2 j_2 0}\left[\frac{1}{(x_{j_1} - a_{j_1 j_1 1})^{m-1}} - \frac{1}{(x_{j_1 0} - a_{j_1 j_1 1})^{m-1}}\right] \\
& = a_{j_1 j_1 0}(x_{j_2 0} - a_{j_1 j_2 1})^n (t - t_0), \\
& \text{for } m \neq 1; \\
& a_{j_2 j_2 0} \ln \frac{|x_{j_1} - a_{j_1 j_1 1}|}{|x_{j_1 0} - a_{j_1 j_1 1}|} = a_{j_1 j_1 0}(x_{j_2 0} - a_{j_1 j_2 1})^n (t - t_0), \\
& \text{for } m = 1.
\end{aligned}
$$

(i_1) At $(x_{j_1}^*, x_{j_2 0}) = (a_{j_1 j_1 1}, x_{j_2 0})$ and $x_{j_2 0} \neq a_{j_1 j_2 1}$, the variational equation is

$$
\begin{aligned}
x_{j_2} &= x_{j_2 0}, \\
\Delta \dot{x}_{j_1} &= a_{j_1 j_1 0}(x_{j_2 0} - a_{j_1 j_2 1})^n (\Delta x_{j_1})^m.
\end{aligned}
$$

The flow at $(x_{j_1}^*, x_{j_2 0}) = (a_{j_1 j_1 1}, x_{j_2 0})$ are the $(2m_1 + 1)^{\text{th}}$-order source and sink for $a_{j_1 j_1 0}(x_{j_2 0} - a_{j_1 j_2 1})^n > 0$ and $a_{j_1 j_1 0}(x_{j_2 0} - a_{j_1 j_2 1})^n < 0$ for $m = 2m_1 + 1$. The flow at $(x_{j_1}^*, x_{j_2 0}) = (a_{j_1 j_1 1}, x_{j_2 0})$ are the $(2m_1)^{\text{th}}$-order upper-saddle and lower-saddle for $a_{j_1 j_1 0}(x_{j_2 0} - a_{j_1 j_2 1})^n > 0$ and $a_{j_1 j_1 0}(x_{j_2 0} - a_{j_1 j_2 1})^n < 0$ for $m = 2m_1$.

Therefore, at $(x_{j_1}^*, x_{j_2 0}) = (a_{j_1 j_1 1}, x_{j_2 0})$ and $x_{j_2 0} \neq a_{j_1 j_2 1}$, there are two cases $m = 2m_1 + 1$ and $m = 2m_1$. For $m = 2m_1$,
(i_{1a}) For $m = 2m_1 + 1$, the equilibrium of $(x_{j_1}^*, x_{j_2 0}) = (a_{j_1 j_1 1}, x_{j_2 0})$ and $x_{j_2 0} \neq a_{j_1 j_2 1}$ for the x_{j_1}-direction has the following properties as in Eqs. (1.3) and (1.4).

- For $a_{j_1 j_1 0}(x_{j_2 0} - a_{j_1 j_2 1})^n > 0$ and $a_{j_2 j_2 0} = 0$,

$$
(a_{j_1 j_1 1}, x_{j_2 0}) = \underbrace{((2m_1 + 1)^{\text{th}}\text{SO, IC})}_{(2m_1+1)^{th}\text{-order source}}.
$$

The flow at $(x_{j_1}^*, x_{j_2 0}) = (a_{j_1 j_1 1}, x_{j_2 0})$ is a $((2m_1 + 1)^{\text{th}}$SO, IC)-source.

- For $a_{j_1 j_1 0}(x_{j_2 0} - a_{j_1 j_2 1})^n < 0$ and $a_{j_2 j_2 0} = 0$

$$
(a_{j_1 j_1 1}, x_{j_2 0}) = \underbrace{((2m_1 + 1)^{\text{th}}\text{SI, IC})}_{(2m_1+1)^{\text{th}}\text{-order sink}}.
$$

The flow at $(x_{j_1}^*, x_{j_2 0}) = (a_{j_1 j_1 1}, x_{j_2 0})$ is a $((2m_1 + 1)^{\text{th}}$SI, IC)-sink.

(i_{1b}) For $m = 2m_1$, the equilibrium of $(x_{j_1}^*, x_{j_2 0}) = (a_{j_1 j_1 1}, x_{j_2 0})$ and $x_{j_2 0} \neq a_{j_1 j_2 1}$ for the x_{j_1}-direction has the following properties as in Eqs. (1.5) and (1.6).

- For $a_{j_1 j_1 0}(x_{j_2 0} - a_{j_1 j_2 1})^n > 0$ and $a_{j_2 j_2 0} = 0$,

$$
(a_{j_1 j_1 1}, x_{j_2 0}) = \underbrace{((2m_1)^{\text{th}}\text{US, IC})}_{(2m_1)^{\text{th}}\text{-order upper-saddle}}.
$$

The flow at $(x_{j_1}^*, x_{j_20}) = (a_{j_1j_11}, x_{j_20})$ is a $((2m_1)^{\text{th}}$US, IC)-upper-saddle.

- For $a_{j_1j_10}(x_{j_20} - a_{j_1j_21})^n < 0$ and $a_{j_2j_20} = 0$

$$(a_{j_1j_11}, x_{j_20}) = \underbrace{((2m_1)^{\text{th}}\text{LS,IC})}_{(2m_1+1)^{\text{th}}\text{-order lower saddle}}.$$

The flow at $(x_{j_1}^*, x_{j_20}) = (a_{j_1j_11}, x_{j_20})$ is a $((2m_1)^{\text{th}}$LS, IC)-lower-saddle.

(i_2) At $(x_{j_1}, x_{j_2}^*) = (\overline{x}_{j_1}, a_{j_1j_21})$ and $\overline{x}_{j_1} \neq a_{j_1j_11}$, the variational equation is

$$\begin{aligned} x_{j_2} &= x_{j_20}, \\ \Delta\dot{x}_{j_1} &= a_{j_1j_10}(\overline{x}_{j_1} - a_{j_1j_11})^m(\Delta x_{j_1})^n. \end{aligned}$$

The flows at $(x_{j_1}, x_{j_2}^*) = (\overline{x}_{j_1}, a_{j_1j_21})$ and $\overline{x}_{j_1} \neq a_{j_1j_11}$ is of the $(2n_1+1)^{\text{th}}$-order increasing and decreasing-inflection parallel flows for $a_{j_1j_10}(\overline{x}_{j_1} - a_{j_1j_11})^m > 0$ and $a_{j_1j_10}(\overline{x}_{j_1} - a_{j_1j_11})^m < 0$ for $n = 2n_1 + 1$. The flows at $(x_{j_1}, x_{j_2}^*) = (\overline{x}_{j_1}, a_{j_1j_21})$ and $\overline{x}_{j_1} \neq a_{j_1j_11}$ is of the $(2n_1)^{\text{th}}$-order parabola parallel flows for $a_{j_1j_10}(\overline{x}_{j_1} - a_{j_1j_11})^m > 0$ and $a_{j_1j_10}(\overline{x}_{j_1} - a_{j_1j_11})^m < 0$ for $n = 2n_1$.

Therefore, at $(x_{j_1}, x_{j_2}^*) = (\overline{x}_{j_1}, a_{j_1j_21})$ and $\overline{x}_{j_1} \neq a_{j_1j_11}$, there are two cases $n = 2n_1 + 1$ and $n = 2n_1$.

(i_{2a}) For $n = 2n_1 + 1$, the equilibrium of $(x_{j_1}, x_{j_2}^*) = (\overline{x}_{j_1}, a_{j_1j_21})$ and $\overline{x}_{j_1} \neq a_{j_1j_11}$ has the following properties as in Eqs. (1.7) and (1.8).

- For $a_{j_1j_10}(\overline{x}_{j_1} - a_{j_1j_11})^m > 0$ and $a_{j_2j_20} = 0$,

$$(\overline{x}_{j_1}, a_{j_1j_21}) = \underbrace{(\text{PL}, (2\text{n}_1+1)^{\text{th}}\text{II})}_{(2n_1+1)^{\text{th}}\text{order increasing-inflection parallel flow}}.$$

The flow at $(x_{j_1}, x_{j_2}^*) = (\overline{x}_{j_1}, a_{j_1j_21})$ and $\overline{x}_{j_1} \neq a_{j_1j_11}$ is a (PL, $(2\text{n}_1+1)^{\text{th}}$II)-increasing-inflection parallel flow.

- For $a_{j_1j_10}(\overline{x}_{j_1} - a_{j_1j_11})^m < 0$ and $a_{j_2j_20} = 0$,

$$(\overline{x}_{j_1}, a_{j_1j_21}) = \underbrace{(\text{PL}, (2\text{n}_1+1)^{\text{th}}\text{DI})}_{(2n_1+1)^{\text{th}}\text{order decreasing-inflection parallel flow}}.$$

The flow at $(x_{j_1}, x_{j_2}^*) = (\overline{x}_{j_1}, a_{j_1j_21})$ and $\overline{x}_{j_1} \neq a_{j_1j_11}$ is a (PL, $(2\text{n}_1+1)^{\text{th}}$DI)-increasing-inflection parallel flow.

(i_{2b}) For $n = 2n_1$, the equilibrium of $(x_{j_1}, x_{j_2}^*) = (\overline{x}_{j_1}, a_{j_1j_2 1})$ and $\overline{x}_{j_1} \neq a_{j_1j_1 1}$ has the following properties as in Eqs. (1.9) and (1.10).

- For $a_{j_1j_1 0}(\overline{x}_{j_1} - a_{j_1j_1 1})^m > 0$ and $a_{j_2j_2 0} = 0$,

$$(\overline{x}_{j_1}, a_{j_1j_2 1}) = \underbrace{(\mathrm{PL}, (2\mathrm{n}_1)^{\mathrm{th}}\mathrm{UP})}_{(2n_1)^{\mathrm{th}}\text{order up-parabola parallel flow}}.$$

The flow at $(x_{j_1}, x_{j_2}^*) = (\overline{x}_{j_1}, a_{j_1j_2 1})$ and $\overline{x}_{j_1} \neq a_{j_1j_1 1}$ is a (PL, $(2\mathrm{n}_1)^{\mathrm{th}}$UP)-up-parabola parallel flow.

- For $a_{j_1j_1 0}(\overline{x}_{j_1} - a_{j_1j_1 1})^m < 0$ and $a_{j_2j_2 0} = 0$,

$$(\overline{x}_{j_1}, a_{j_1j_2 1}) = \underbrace{(\mathrm{PL}, (2\mathrm{n}_1)^{\mathrm{th}}\mathrm{DP})}_{(2n_1)^{\mathrm{th}}\text{order down-parabola parallel flow}}.$$

The flow at $(x_{j_1}, x_{j_2}^*) = (\overline{x}_{j_1}, a_{j_1j_2 1})$ and $\overline{x}_{j_1} \neq a_{j_1j_1 1}$ is a (PL, $(2n_1)^{\mathrm{th}}$DP)-down-parabola parallel flow.

(i_3) From cases (i_1) and (i_2), let $\overline{x}_{j_1} = a_{j_1j_1 1}$ and $x_{j_2 0} = a_{j_1j_2 1}$. At $(x_{j_1}^*, x_{j_2}^*) = (a_{j_1j_1 1},$, $a_{j_1j_2 1})$, *there are four cases*: $m = 2m_1 + 1$ and $n = 2n_1 + 1$, $m = 2m_1 + 1$ and $n = 2n_1$; $m = 2m_1$ and $n = 2n_1 + 1$; and $m = 2m_1$ and $n = 2n_1$.

(i_{3a}) For $m = 2m_1 + 1$ and $n = 2n_1 + 1$, the equilibrium of $(x_{j_1}^*, x_{j_2}^*) = (a_{j_1j_1 1}, a_{j_1j_2 1})$ has the following properties as in Eqs. (1.11) and (1.12)

- For $a_{j_1j_1 0} > 0$ and $a_{j_2j_2 0} = 0$,

$$(a_{j_1j_1 1}, a_{j_1j_2 1}) = \underbrace{((2m_1 + 1)^{\mathrm{th}}\mathrm{SI{:}SO}, (2n_1 + 1)^{\mathrm{th}}\mathrm{DI{:}II})}_{[(2m_1+1),(2n_1+1)]\text{-inflection sink-to-source switch}}.$$

The flow at $(x_{j_1}^*, x_{j_2}^*) = (a_{j_1j_1 1}, a_{j_1j_2 1})$ is a ($(2m_1 + 1)^{\mathrm{th}}$SI:SO,$(2n_1 + 1)^{\mathrm{th}}$DI : II)-inflection sink-to-source switch.

- For $a_{j_1j_1 0} < 0$ and $a_{j_2j_2 0} = 0$,

$$(a_{j_1j_1 1}, a_{j_1j_2 1}) = \underbrace{((2m_1 + 1)^{\mathrm{th}}\mathrm{SO{:}SI}, (2n_1 + 1)^{\mathrm{th}}\mathrm{II{:}DI})}_{[(2m_1+1),(2n_1+1)]\text{-inflection source-to-sink switch}}.$$

The flow at $(x_{j_1}, x_{j_2}^*) = (\overline{x}_{j_1}, a_{j_1j_2 1})$ is a ($(2m_1 + 1)^{\mathrm{th}}$SI:SO,$(2n_1 + 1)^{\mathrm{th}}$DI : II)-inflection source-to-sink switch.

(i_{3b}) For $m = 2m_1 + 1$ and $n = 2n_1$, the equilibrium of $(x_{j_1}^*, x_{j_2}^*) = (a_{j_1j_1 1}, a_{j_1j_2 1})$ has the following properties as in Eqs. (1.13) and (1.14).

- For $a_{j_1j_10} > 0$ and $a_{j_2j_20} = 0$,

$$(a_{j_1j_11}, a_{j_1j_21}) = \underbrace{((2m_1+1)^{\text{th}}\text{SO:SO},(2n_1)^{\text{th}}\text{DP:UP})}_{[(2m_1+1),(2n_1)]\text{-parabola sink-to-source switch}}.$$

The flow at $(x^*_{j_1}, x^*_{j_2}) = (a_{j_1j_11}, a_{j_1j_21})$ is a $((2m_1+1)^{\text{th}}\text{SO:SO},(2n_1+1)^{\text{th}}\text{DI}:\text{II})$-inflection source-to-source switch.

- For $a_{j_1j_10} < 0$ and $a_{j_2j_20} = 0$,

$$(a_{j_1j_11}, a_{j_1j_21}) = \underbrace{((2m_1+1)^{\text{th}}\text{SI:SI},(2n_1)^{\text{th}}\text{DP:UP})}_{[(2m_1+1),(2n_1)]\text{-parabola sink-to-sink switch}}.$$

The flow at $(x_{j_1}, x^*_{j_2}) = (\overline{x}_{j_1}, a_{j_1j_21})$ is a $((2m_1+1)^{\text{th}}\text{SI:SI},(2n_1+1)^{\text{th}}\text{DI}:\text{II})$-inflection sink-to-sink switch.

($\text{i}_{3\text{c}}$) For $m = 2m_1$ and $n = 2n_1 + 1$, the equilibrium of $(x^*_{j_1}, x^*_{j_2}) = (a_{j_1j_11}, a_{j_1j_21})$ has the following properties as in Eqs. (1.15) and (1.16).

- For $a_{j_1j_10} > 0$ and $a_{j_2j_20} = 0$,

$$(a_{j_1j_11}, a_{j_1j_21}) = \underbrace{((2m_1)^{\text{th}}\text{LS:US},(2n_1+1)^{\text{th}}\text{II:II})}_{[(2m_1),(2n_1+1)]\text{-inflection lower-to-upper saddle switch}}.$$

The flow at $(x^*_{j_1}, x^*_{j_2}) = (a_{j_1j_11}, a_{j_1j_21})$ is a $((2m_1)^{\text{th}}\text{LS:US},(2n_1+1)^{\text{th}}\text{II}:\text{II})$-inflection lower-to-upper saddle switch.

- For $a_{j_1j_10} < 0$ and $a_{j_2j_20} = 0$,

$$(a_{j_1j_11}, a_{j_1j_21}) = \underbrace{((2m_1)^{\text{th}}\text{US:LS},(2n_1+1)^{\text{th}}\text{DI:DI})}_{[(2m_1),(2n_1+1)]\text{-inflection upper-to-lower saddle switch}}.$$

The flow at $(x_{j_1}, x^*_{j_2}) = (\overline{x}_{j_1}, a_{j_1j_21})$ is a $((2m_1)^{\text{th}}\text{US:LS},(2n_1+1)^{\text{th}}\text{DI}:\text{DI})$-inflection upper-to-lower saddle switch.

($\text{i}_{3\text{d}}$) For $m = 2m_1$ and $n = 2n_1$, the equilibrium of $(x^*_{j_1}, x^*_{j_2}) = (a_{j_1j_11}, a_{j_1j_21})$ has the following properties as in Eqs. (1.17) and (1.18).

- For $a_{j_1j_10} > 0$ and $a_{j_2j_20} = 0$,

$$(a_{j_1j_11}, a_{j_1j_21}) = \underbrace{((2m_1)^{\text{th}}\text{US:US},(2n_1)^{\text{th}}\text{UP:UP})}_{[(2m_1),(2n_1)]\text{-parabola upper-to-upper saddle switch}}.$$

The flow at $(x^*_{j_1}, x^*_{j_2}) = (a_{j_1j_11}, a_{j_1j_21})$ is a $((2m_1)^{\text{th}}$US:US,$(2n_1)^{\text{th}}$UP : UP)-inflection upper-to-upper saddle switch.

- For $a_{j_1j_10} < 0$ and $a_{j_2j_20} = 0$,

$$(a_{j_1j_11}, a_{j_1j_21}) = \underbrace{((2m_1)^{\text{th}}\text{US:US},(2n_1)^{\text{th}}\text{UP:UP})}_{[(2m_1),(2n_1)]\text{-parabola lower-to-lower saddle switch}}.$$

The flow at $(x_{j_1}, x^*_{j_2}) = (\overline{x}_{j_1}, a_{j_1j_21})$ is a $((2m_1)^{\text{th}}$LS:LS,$(2n_1)^{\text{th}}$DP : DP) inflection lower-to-lower saddle switch.

2.2 Hyperbolic Flows with Inflection-Source, Sink, Saddle Flows

(ii) If $a_{j_2j_20} \neq 0$, the 2-dimensional dynamical system does not have any equilibriums for the x_{j_2}-direction on the x_{j_2}-direction. The standard form is

$$\begin{aligned}
\dot{x}_{j_2} &= a_{j_2j_20},\\
\dot{x}_{j_1} &= a_{j_1j_10}(x_{j_1} - a_{j_1j_11})^m (x_{j_2} - a_{j_1j_21})^n.
\end{aligned}$$

The foregoing equation becomes

$$\frac{dx_{j_1}}{dx_{j_2}} = \frac{a_{j_1j_10}(x_{j_1} - a_{j_1j_11})^m (x_{j_2} - a_{j_1j_21})^n}{a_{j_2j_20}},$$

or

$$\frac{a_{j_2j_20}dx_{j_1}}{(x_{j_1} - a_{j_1j_11})^m} = a_{j_1j_10}(x_{j_2} - a_{j_1j_21})^n dx_{j_2}.$$

With an initial condition (x_{j_10}, x_{j_20}) art time $t = t_0$, integration of the foregoing equation gives the first integral manifold as

$$
\begin{aligned}
&-\frac{1}{m-1}a_{j_2j_20}\Big[\frac{1}{(x_{j_1}-a_{j_1j_11})^{m-1}}-\frac{1}{(x_{j_10}-a_{j_1j_11})^{m-1}}\Big]\\
&=\frac{1}{n+1}a_{j_1j_10}[(x_{j_2}-a_{j_1j_21})^{n+1}-(x_{j_20}-a_{j_1j_21})^{n+1}],\\
&\text{for } m\neq 1;\\
&a_{j_2j_20}\ln\frac{|x_{j_1}\quad a_{j_1j_11}|}{|x_{j_10}-a_{j_1j_11}|}=\tfrac{1}{n+1}a_{j_1j_10}[(x_{j_2}-a_{j_1j_21})^{n+1}-(x_{j_20}-a_{j_1j_21})^{n+1}],\\
&\text{for } m=1.
\end{aligned}
$$

At the equilibrium of $x^*_{j_2}=a_{j_1j_21}$ and $\overline{x}_{j_1}\neq a_{j_1j_11}$, in phase space,

$$
\begin{aligned}
&\frac{dx_{j_1}}{dx_{j_2}}\Big|_{x^*_{j_2}=a_{j_1j_21}}=\frac{a_{j_1j_10}(\overline{x}_{j_1}-a_{j_1j_11})^m(x_{j_2}-a_{j_1j_21})^n}{a_{j_2j_20}}\Big|_{x^*_{j_2}=a_{j_1j_21}}=0,\\
&\vdots\\
&\frac{d^nx_{j_1}}{dx^n_{j_2}}\Big|_{x^*_{j_2}=a_{j_1j_21}}=0.
\end{aligned}
$$

If

$$
\frac{d^{n+1}x_{j_1}}{dx^{n+1}_{j_2}}\Big|_{x^*_{j_2}=a_{j_1j_21}}=\frac{a_{j_1j_10}(\overline{x}_{j_1}-a_{j_1j_11})^m n!}{a_{j_2j_20}}>0,
$$

the $(2n_1+1)^{\text{th}}$-order up-parabola flows exists for $n=2n_1+1$, and the $(2n_1)^{\text{th}}$-order increasing-inflection flow exists for $n=2n_1$. If

$$
\frac{d^{n+1}x_{j_1}}{dx^{n+1}_{j_2}}\Big|_{x^*_{j_2}=a_{j_1j_21}}=\frac{a_{j_1j_10}(\overline{x}_{j_1}-a_{j_1j_11})^m n!}{a_{j_2j_20}}<0,
$$

the $(2n_1+1)^{\text{th}}$-order down-parabola flows exists for $n=2n_1+1$, and the $(2n_1)^{\text{th}}$-order decreasing-inflection flow exists for $n=2n_1$.

At the equilibrium of $x^*_{j_1}=a_{j_1j_11}$ and $\overline{x}_{j_2}\neq a_{j_1j_21}$, the variational equation is

$$
\begin{aligned}
\dot{x}_{j_2}&=a_{j_2j_20},\\
\Delta\dot{x}_{j_1}&=a_{j_1j_10}(\overline{x}_{j_2}-a_{j_1j_21})^n(\Delta x_{j_1})^m.
\end{aligned}
$$

The flows at $x^*_{j_1}=a_{j_1j_11}$ and $\overline{x}_{j_2}\neq a_{j_1j_21}$ are the $(2m_1+1)^{\text{th}}$-order source and sink for $a_{j_1j_10}(\overline{x}_{j_2}-a_{j_1j_21})^n>0$ and $a_{j_1j_10}(\overline{x}_{j_2}-a_{j_1j_21})^n<0$ for $m=2m_1+1$. The flows at $x^*_{j_1}=a_{j_1j_11}$ and $\overline{x}_{j_2}\neq a_{j_1j_21}$ are the $(2m_1)^{\text{th}}$-order upper-saddle and lower-saddle for $a_{j_1j_10}(\overline{x}_{j_2}-a_{j_1j_21})^n>0$ and $a_{j_1j_10}(\overline{x}_{j_2}-a_{j_1j_21})^n<0$ for $m=2m_1$.

Let $\overline{x}_{j_1} = a_{j_1 j_1 1}$ and $\overline{x}_{j_2} = a_{j_1 j_2 1}$. Therefore, from the above discussion, there are four cases: (ii$_1$) $m = 2m_1 + 1$ and $n = 2n_1 + 1$, (ii$_2$) $m = 2m_1$ and $n = 2n_1 + 1$, (ii$_3$) $m = 2m_1 + 1$ and $n = 2n_1$,(ii$_4$) $m = 2m_1$ and $n = 2n_1$.

(ii$_1$) For $m = 2m_1 + 1$ and $n = 2n_1 + 1$, the equilibrium of $(x^*_{j_1}, x^*_{j_2}) = (a_{j_1 j_1 1}, a_{j_1 j_2 1})$ for the x_{j_1}-direction has the following properties as in Eqs. (1.21–1.24).

- For $a_{j_1 j_1 0} > 0$ and $a_{j_2 j_2 0} > 0$,

$$(a_{j_1 j_1 1}, a_{j_1 j_2 1}) = \underbrace{([(2m_1{+}1), (2n_1 + 1)]\text{DP:UP,pF})}_{[(2m_1+1),(2n_1+1)]\text{-hyperbolic flow }(+)}.$$

The flow at $(x^*_{j_1}, x^*_{j_2}) = (a_{j_1 j_1 1}, a_{j_1 j_2 1})$ is a ([(2m_1+1), (2n_1 + 1)]DP:UP,pF)-positive hyperbolic flow. The line of $x^*_{j_1} = a_{j_1 j_1 1}$ is a $(2m_1{+}1)^{\text{th}}$-order positive sink-to-source directrix flow.

- For $a_{j_1 j_1 0} < 0$ and $a_{j_2 j_2 0} > 0$,

$$(a_{j_1 j_1 1}, a_{j_1 j_2 1}) = \underbrace{([(2m_1{+}1), (2n_1 + 1)]\text{UP:DP,pF})}_{[(2m_1+1),(2n_1+1)]\text{-hyperbolic-secant flow }(+)}.$$

The flow at $(x^*_{j_1}, x^*_{j_2}) = (a_{j_1 j_1 1}, a_{j_1 j_2 1})$ is a ([(2m_1+1), (2n_1 + 1)]UP:DP,pF)-positive hyperbolic-secant flow. The line of $x^*_{j_1} = a_{j_1 j_1 1}$ is a $(2m_1{+}1)^{\text{th}}$-order positive source-to-sink directrix flow.

- For $a_{j_1 j_1 0} > 0$ and $a_{j_2 j_2 0} < 0$,

$$(a_{j_1 j_1 1}, a_{j_1 j_2 1}) = \underbrace{([(2m_1{+}1), (2n_1 + 1)]\text{UP:DP,nF})}_{[(2m_1+1),(2n_1+1)]\text{-hyperbolic-secant flow }(-)}.$$

The flow at $(x^*_{j_1}, x^*_{j_2}) = (a_{j_1 j_1 1}, a_{j_1 j_2 1})$ is a ([(2m_1+1), (2n_1 + 1)]UP:DP,nF)-negative hyperbolic-secant flow. The line of $x^*_{j_1} = a_{j_1 j_1 1}$is a $(2m_1{+}1)^{\text{th}}$-order negative sink-to-source directrix flow.

- For $a_{j_1 j_1 0} < 0$ and $a_{j_2 j_2 0} < 0$,

$$(a_{j_1 j_1 1}, a_{j_1 j_2 1}) = \underbrace{([(2m_1{+}1), (2n_1 + 1)]\text{UP:DP,nF})}_{[(2m_1+1),(2n_1+1)]\text{-hyperbolic-secant flow }(-)}.$$

The flow at $(x^*_{j_1}, x^*_{j_2}) = (a_{j_1 j_1 1}, a_{j_1 j_2 1})$ is a ([(2m_1+1), (2n_1 + 1)]UP:DP,nF)-negative hyperbolic flow. The line of $x^*_{j_1} = a_{j_1 j_1 1}$ is a $(2m_1{+}1)^{\text{th}}$-order negative source-to-sink directrix flow.

(ii$_2$) For $m = 2m_1$ and $n = 2n_1 + 1$, the equilibrium of $(x^*_{j_1}, x^*_{j_2}) = (a_{j_1j_1 1}, a_{j_1j_2 1})$ for the x_{j_1}-direction has the following properties as in Eqs. (1.25–1.28).

- For $a_{j_1j_1 0} > 0$ and $a_{j_2j_2 0} > 0$,

$$(a_{j_1j_1 1}, a_{j_1j_2 1}) = \underbrace{([(2m_1), (2n_1+1)]\text{UP:UP,pF})}_{[(2m_1),(2n_1+1)]\text{-hyperbolic-secant-to-hyperbolic flow }(+)}.$$

The flow at $(x^*_{j_1}, x^*_{j_2}) = (a_{j_1j_1 1}, a_{j_1j_2 1})$ is a ([(2m_1), (2n_1 + 1)]UP:UP,pF)-positive hyperbolic-secant to hyperbolic flow. The line of $x^*_{j_1} = a_{j_1j_1 1}$ is a $(2m_1)^{\text{th}}$-order positive lower-to-upper saddle directrix flow.

- For $a_{j_1j_1 0} < 0$ and $a_{j_2j_2 0} > 0$,

$$(a_{j_1j_1 1}, a_{j_1j_2 1}) = \underbrace{([(2m_1), (2n_1+1)]\text{DP:DP,pF})}_{[(2m_1),(2n_1+1)]\text{-hyperbolic-to-hyperbolic-secant flow }(+)}.$$

The flow at $(x^*_{j_1}, x^*_{j_2}) = (a_{j_1j_1 1}, a_{j_1j_2 1})$ is a ([(2m_1), (2n_1 + 1)]DP:DP,pF)-positive hyperbolic to hyperbolic-secant flow. The line of $x^*_{j_1} = a_{j_1j_1 1}$ is a $(2m_1)^{\text{th}}$-order positive upper-to-lower saddle directrix flow.

- For $a_{j_1j_1 0} > 0$ and $a_{j_2j_2 0} < 0$,

$$(a_{j_1j_1 1}, a_{j_1j_2 1}) = \underbrace{([(2m_1), (2n_1+1)]\text{DP:DP,nF})}_{[(2m_1),(2n_1+1)]\text{-hyperbolic-to-hyperbolic-secant flow }(-)}.$$

The flow at $(x^*_{j_1}, x^*_{j_2}) = (a_{j_1j_1 1}, a_{j_1j_2 1})$ is a ([(2m_1), (2n_1 + 1)]DP:DP,nF)-negative hyperbolic-to-hyperbolic-secant flow. The line of $x^*_{j_1} = a_{j_1j_1 1}$ is a $(2m_1)^{\text{th}}$-order negative lower-to-upper saddle directrix flow.

- For $a_{j_1j_1 0} < 0$ and $a_{j_2j_2 0} < 0$,

$$(a_{j_1j_1 1}, a_{j_1j_2 1}) = \underbrace{([(2m_1), (2n_1+1)]\text{UP:UP,nF})}_{[(2m_1),(2n_1+1)]\text{-hyperbolic-secant-to-hyperbolic flow }(-)}.$$

The flow at $(x^*_{j_1}, x^*_{j_2}) = (a_{j_1j_1 1}, a_{j_1j_2 1})$ is a ([(2m_1), (2n_1 + 1)]UP:UP,nF)-negative hyperbolic-secant-to-hyperbolic flow. The line of $x^*_{j_1} = a_{j_1j_1 1}$ is a $(2m_1)^{\text{th}}$-order negative upper-to-lower saddle directrix flow.

(ii$_3$) For $m = 2m_1 + 1$ and $n = 2n_1$, the equilibrium of $(x^*_{j_1}, x^*_{j_2}) = (a_{j_1j_1 1}, a_{j_1j_2 1})$ for the x_{j_1}-direction has the following properties as in Eqs. (1.29–1.32).

- For $a_{j_1j_10} > 0$ and $a_{j_2j_20} > 0$,

$$(a_{j_1j_11}, a_{j_1j_21}) = \underbrace{([(2m_1{+}1), (2n_1)]_{\text{DI:II}}\text{SO,pF})}_{[(2m_1+1),(2n_1)]\text{-decreasing-to-increasing inflection source flow}(+)}.$$

The flow at $(x^*_{j_1}, x^*_{j_2}) = (a_{j_1j_11}, a_{j_1j_21})$ is a $([(2m_1{+}1), (2n_1)]_{\text{DI:II}}\text{SO,pF})$-positive decreasing-to-increasing inflection source flow. The line of $x^*_{j_1} = a_{j_1j_11}$ is a $(2m_1{+}1)^{\text{th}}$-order positive source directrix flow.

- For $a_{j_1j_10} < 0$ and $a_{j_2j_20} > 0$,

$$(a_{j_1j_11}, a_{j_1j_21}) = \underbrace{([(2m_1{+}1), (2n_1)]_{\text{II:DI}}\text{SI,pF})}_{[(2m_1+1),(2n_1)]\text{-increasing-to-decreasing inflection sink flow}(+)}.$$

The flow at $(x^*_{j_1}, x^*_{j_2}) = (a_{j_1j_11}, a_{j_1j_21})$ is a $([(2m_1{+}1), (2n_1)]_{\text{II:DI}}\text{SI,pF})$-positive increasing-to-decreasing inflection source flow. The line of $x^*_{j_1} = a_{j_1j_11}$ is a $(2m_1{+}1)^{\text{th}}$-order positive sink directrix flow.

- For $a_{j_1j_10} > 0$ and $a_{j_2j_20} < 0$,

$$(a_{j_1j_11}, a_{j_1j_21}) = \underbrace{([(2m_1{+}1), (2n_1)]_{\text{II:DI}}\text{SO,nF})}_{[(2m_1+1),(2n_1)]\text{-increasing-to-decreasing inflection source flow}(-)}.$$

The flow at $(x^*_{j_1}, x^*_{j_2}) = (a_{j_1j_11}, a_{j_1j_21})$ is a $([(2m_1{+}1), (2n_1)]_{\text{II:DI}}\text{SO,nF})$-negative increasing-to-decreasing inflection source flow. The line of $x^*_{j_1} = a_{j_1j_11}$ is a $(2m_1{+}1)^{\text{th}}$-order negative source directrix flow.

- For $a_{j_1j_10} < 0$ and $a_{j_2j_20} < 0$,

$$(a_{j_1j_11}, a_{j_1j_21}) = \underbrace{([(2m_1{+}1), (2n_1)]_{\text{DI:II}}\text{SI,nF})}_{[(2m_1+1),(2n_1)]\text{-decreasing-to-increasing inflection sink flow}(-)}.$$

The flow at $(x^*_{j_1}, x^*_{j_2}) = (a_{j_1j_11}, a_{j_1j_21})$ is a $([(2m_1{+}1), (2n_1)]_{\text{DI:II}}\text{SI,nF})$-negative decreasing-to-increasing inflection sink flow. The line of $x^*_{j_1} = a_{j_1j_11}$ is a $(2m_1{+}1)^{\text{th}}$-order negative sink directrix flow.

(ii_4) For $m = 2m_1$ and $n = 2n_1$, the equilibrium of $(x^*_{j_1}, x^*_{j_2}) = (a_{j_1j_11}, a_{j_1j_21})$ for the x_{j_1}-direction has the following properties as in Eqs. (1.33–1.36).

- For $a_{j_1j_10} > 0$ and $a_{j_2j_20} > 0$,

$$(a_{j_1j_11}, a_{j_1j_21}) = \underbrace{([(2m_1), (2n_1)]_{\text{II:II}}\text{US,pF})}_{[(2m_1),(2n_1)]\text{-increasing-inflection upper-saddle flow }(+)}.$$

The flow at $(x_{j_1}^*, x_{j_2}^*) = (a_{j_1j_1}1, a_{j_1j_2}1)$ is a $([(2m_1), (2n_1)]_{\text{II:II}}\text{US,pF})$-positive increasing-inflection upper-saddle flow. The line of $x_{j_1}^* = a_{j_1j_1}1$ is a $(2m_1)^{\text{th}}$-order positive upper-saddle directrix flow.

- For $a_{j_1j_1}0 < 0$ and $a_{j_2j_2}0 > 0$,

$$(a_{j_1j_1}1, a_{j_1j_2}1) = \underbrace{([(2m_1), (2n_1)]_{\text{DI:DI}}\text{LS,pF})}_{[(2m_1),(2n_1)]\text{-decreasing-inflection lower-saddle flow }(+)}.$$

The flow at $(x_{j_1}^*, x_{j_2}^*) = (a_{j_1j_1}1, a_{j_1j_2}1)$ is a $([(2m_1), (2n_1)]_{\text{DI:DI}}\text{LS,pF})$-positive decreasing-inflection lower-saddle flow. The line of $x_{j_1}^* = a_{j_1j_1}1$ is a $(2m_1)^{\text{th}}$-order positive lower-saddle directrix flow.

- For $a_{j_1j_1}0 > 0$ and $a_{j_2j_2}0 < 0$,

$$(a_{j_1j_1}1, a_{j_1j_2}1) = \underbrace{([(2m_1), (2n_1)]_{\text{DI:DI}}\text{US,nF})}_{[(2m_1),(2n_1)]\text{-decreasing-inflection upper-saddle flow }(-)}.$$

The flow at $(x_{j_1}^*, x_{j_2}^*) = (a_{j_1j_1}1, a_{j_1j_2}1)$ is a $([(2m_1), (2n_1)]_{\text{DI:DI}}\text{US,nF})$-negative decreasing-inflection upper-saddle flow. The line of $x_{j_1}^* = a_{j_1j_1}1$ is a$(2m_1)^{\text{th}}$-order negative upper-saddle directrix flow.

- For $a_{j_1j_1}0 < 0$ and $a_{j_2j_2}0 < 0$,

$$(a_{j_1j_1}1, a_{j_1j_2}1) = \underbrace{([(2m_1), (2n_1)]_{\text{II:II}}\text{LS,nF})}_{[(2m_1),(2n_1)]\text{-increasing-inflection lower-saddle flow }(-)}.$$

The flow at $(x_{j_1}^*, x_{j_2}^*) = (a_{j_1j_1}1, a_{j_1j_2}1)$ is a $([(2m_1), (2n_1)]_{\text{II:II}}\text{LS,nF})$-negative increasing-inflection lower-saddle flow. The line of $x_{j_1}^* = a_{j_1j_1}1$ is a $(2m_1)^{\text{th}}$-order positive lower-saddle directrix flow.

2.3 Singular Flows Networks

(iii) Consider

$$\begin{aligned}
\dot{x}_{j_2} &= a_{j_2j_2}0 \\
\dot{x}_{j_1} &= a_{j_1j_1}0 \prod_{\alpha=1}^{p} (x_{j_1} - a_{j_1j_1\alpha})^{m_\alpha} \prod_{\beta=1}^{q} (x_{j_2} - a_{j_1j_2\beta})^{n_\beta},
\end{aligned}$$

where

$$\sum_{\alpha=1}^{p} m_\alpha = m, \sum_{\beta=1}^{q} n_\beta = n.$$

The foregoing equation becomes

$$\frac{dx_{j_1}}{dx_{j_2}} = \frac{a_{j_1j_10}\prod_{\alpha=1}^{p}(x_{j_1}-a_{j_1j_1\alpha})^{m_\alpha}\prod_{\beta=1}^{q}(x_{j_2}-a_{j_1j_2\beta})^{n_\beta}}{a_{j_2j_20}},$$

The deformation gives

$$\frac{a_{j_2j_20}dx_{j_1}}{\prod\limits_{\alpha=1}^{p}(x_{j_1}-a_{j_1j_1\alpha})^{m_\alpha}} = a_{j_1j_10}\prod_{\beta=1}^{q}(x_{j_2}-a_{j_1j_2\beta})^{n_\beta}dx_{j_2},$$

or

$$\frac{a_{j_2j_20}dx_{j_1}}{(x_{j_1}-a_{j_1j_1\alpha_1})^{m_{\alpha_1}}\prod_{\alpha_2=1,\alpha_2\neq\alpha_1}^{p}(x_{j_1}-a_{j_1j_1\alpha_2})^{m_\alpha}}$$
$$= a_{j_1j_10}(x_{j_2} \quad - a_{j_1j_2\beta_1})^{n_{\beta_1}}\prod_{\beta_2=1,\beta_2\neq\beta_1}^{q}(x_{j_2}-a_{j_1j_2\beta_1}$$
$$+ a_{j_1j_2\beta_1}-a_{j_1j_2\beta})^{n_{\beta_2}}dx_{j_2}$$

and

$$a_{j_2j_20}\sum_{\alpha_1=1}^{p}\sum_{l_1=1}^{m_{\alpha_1}}\frac{A_{l_1m_{\alpha_1}}}{(x_{j_1}-a_{j_1j_1\alpha_1})^{l_1}}dx_{j_1}$$
$$= a_{j_1j_10}[(x_{j_2}-a_{j_1j_2\beta_1})^n + \sum_{\rho=1}^{n-n_{\beta_1}} b_{j_1j_2\rho}[(x_{j_2}-a_{j_1j_2\beta_1})^{n-\rho}]dx_{j_2},$$

where

$$A_{l_1m_{\alpha_1}} = \frac{1}{(m_{r_1}-l_1)!}\frac{d^{m_{\alpha_1}-l_1}}{dx_{j_1}^{m_{\alpha_1}-l_1}}\frac{1}{\prod_{\alpha_2=1,\alpha_2\neq\alpha_1}^{p}(x_{j_1}-a_{j_1j_1\alpha_2})^{m_{\alpha_2}}}\Big|_{x_{j_1}^*=a_{j_1j_1\alpha_1}},$$

$$A_{m_{\alpha_1}m_{\alpha_1}} = \frac{1}{\prod_{\alpha_2=1,\alpha_2\neq\alpha_1}^{p}(a_{j_1j_1\alpha_1}-a_{j_1j_1\alpha_2})^{m_{\alpha_2}}};$$

$$b_{j_1j_2\rho} = \sum_{l_{\beta_q},t_{\beta_q}=0}^{l_{\beta_q}+t_{\beta_q}=n_{\beta_q}}\cdots\sum_{l_{\beta_2},t_{\beta_2}=0}^{l_{\beta_2}+t_{\beta_2}=n_{\beta_2}}[\prod_{k=\beta_2}^{\beta_q}\frac{n_k!}{l_k!t_k!}(a_{j_1j_2\beta_1}-a_{j_1j_2k})^{l_k}]\delta_{t_{\beta_2}+\ldots+t_{\beta_q}}^{n-\rho},$$

$$b_{j_1j_20} = 1.$$

With an initial condition $(x_{j_1 0}, x_{j_2 0})$ art time $t = t_0$, integration of the foregoing equation gives the first integral manifold as

$$\begin{aligned}
&a_{j_2j_20}\sum_{\alpha_1=1}^{p}\sum_{l_1=1}^{m_{\alpha_1}}\frac{A_{l_1m_{\alpha_1}}}{l_1-1}[\frac{1}{(x_{j_1}-a_{j_1j_1\alpha_1})^{l_1-1}}-\frac{1}{(x_{j_10}-a_{j_1j_1\alpha_1})^{l_1-1}}]\\
&=a_{j_1j_10}\left\{\frac{1}{n+1}\left[(x_{j_2}-a_{j_1j_2\beta_1})^{n+1}-(x_{j_20}-a_{j_1j_2\beta_1})^{n+1}\right]\right.\\
&\left.+\sum_{\rho=1}^{n-n_{\beta_1}}\frac{1}{n-\rho+1}b_{j_1j_2\rho}\left[(x_{j_2}-a_{j_1j_2\beta_1})^{n-\rho+1}-(x_{j_20}-a_{j_1j_2\beta_1})^{n-\rho+1}\right]\right\},\\
&\text{for } l_1\neq 1,\\
&\lim_{l_1\to 1}\frac{1}{l_1-1}[\frac{1}{(x_{j_1}-a_{j_1j_1\alpha_1})^{l_1-1}}-\frac{1}{(x_{j_10}-a_{j_1j_1\alpha_1})^{l_1-1}}]=-\ln\frac{|x_{j_1}-a_{j_1j_1\alpha_1}|}{|x_{j_10}-a_{j_1j_1\alpha_1}|},\\
&\text{for } l_1=1.
\end{aligned}$$

At the equilibrium of $x_{j_2}^* = a_{j_1j_2\beta_1}$ $(\beta_1, \beta_2 \in \{1, 2, \ldots, q\}, \beta_1 \neq \beta_2)$ and $\overline{x}_{j_1} \neq a_{j_1j_1\alpha_1}$ $(\alpha = 1, 2, \ldots, p)$, in phase space,

$$\begin{aligned}
\left.\frac{dx_{j_1}}{dx_{j_2}}\right|_{x_{j_2}^*=a_{j_1j_2\beta_1}} &= \frac{a_{j_1j_10}}{a_{j_2j_20}}\prod_{\alpha=1}^{p}(\overline{x}_{j_1}-a_{j_1j_1\alpha})^{m_\alpha}\\
&\times (x_{j_2}-a_{j_1j_2\beta_1})^{n_{\beta_1}}\prod_{\beta_2=1,\beta_2\neq\beta_1}^{q}(x_{j_2}-a_{j_1j_2\beta_2})^{n_{\beta_2}}\Big|_{x_{j_2}^*=a_{j_1j_2\beta_1}}\\
&=0,\\
&\vdots\\
\left.\frac{d^{n_{\beta_1}}x_{j_1}}{dx_{j_2}^{n_{\beta_1}}}\right|_{x_{j_2}^*=a_{j_1j_2\beta_1}} &= 0.
\end{aligned}$$

If

$$\begin{aligned}
\left.\frac{d^{n_{\beta_1}+1}x_{j_1}}{dx_{j_2}^{n_{\beta_1}+1}}\right|_{x_{j_2}^*=a_{j_1j_2\beta_1}} &= \frac{a_{j_1j_10}}{a_{j_2j_20}}n_{\beta_1}!\prod_{\alpha=1}^{p}(\overline{x}_{j_1}-a_{j_1j_1\alpha})^{m_\alpha}\prod_{\beta_2=1,\beta_2\neq\beta_1}^{q}(a_{j_1j_2\beta_1}-a_{j_1j_2\beta_2})^{n_{\beta_2}}\\
&>0,
\end{aligned}$$

the $(2n_{\beta_1}+1)^{\text{th}}$-order up-parabola flow exists for $n_{\beta_1} = 2n_{\beta_1 1}+1$, and the $(2n_{\beta_1 1})^{\text{th}}$—order increasing-inflection flow exists for $n_{\beta_1} = 2n_{\beta_1 1}$. If

$$\begin{aligned}
\left.\frac{d^{n_{\beta_1}+1}x_{j_1}}{dx_{j_2}^{n_{\beta_1}+1}}\right|_{x_{j_2}^*=a_{j_1j_2\beta_1}} &= \frac{a_{j_1j_10}}{a_{j_2j_20}}n_{\beta_1}!\prod_{\alpha=1}^{p}(\overline{x}_{j_1}-a_{j_1j_1\alpha})^{m_\alpha}\prod_{\beta_2=1,\beta_2\neq\beta_1}^{q}(a_{j_1j_2\beta_1}-a_{j_1j_2\beta_2})^{n_{\beta_2}}\\
&<0,
\end{aligned}$$

the $(2n_{\beta_1}+1)^{\text{th}}$-order down-parabola flow exists for $n_{\beta_1} = 2n_{\beta_1 1}+1$, and the $(2n_{\beta_1 1})^{\text{th}}$-order decreasing-inflection flow exists for $n_{\beta_1} = 2n_{\beta_1 1}$.

At the equilibrium of $x_{j_1}^* = a_{j_1 j_1 \alpha_1}$ ($\alpha_1 \in \{1, 2, \ldots, p\}$) and $\overline{x}_{j_2} \neq a_{j_1 j_2 \beta}$ ($\beta \in \{1, 2, \ldots, q\}$) the variational equation is

$$\begin{aligned}\dot{x}_{j_2} &= a_{j_2 j_2 0},\\ \Delta\dot{x}_{j_1} &= a_{j_1 j_1 0} \prod_{\alpha_2=1,\alpha_2\neq\alpha_1}^{p} (a_{j_1 j_1 \alpha_1} - a_{j_1 j_1 \alpha_2})^{m_{\alpha_2}} \prod_{\beta=1}^{q} (\overline{x}_{j_2} - a_{j_1 j_2 \beta})^{n_\beta} (\Delta x_{j_1})^{m_{\alpha_1}}.\end{aligned}$$

The flows at $x_{j_1}^* = a_{j_1 j_1 \alpha_1}$ and $\overline{x}_{j_2} \neq a_{j_1 j_2 \beta}$ are the $(2m_{\alpha_1 1} + 1)^{\text{th}}$-order source and sink for

$$a_{j_1 j_1 0} \prod_{\alpha_2=1,\alpha_2\neq\alpha_1}^{p} (a_{j_1 j_1 \alpha_1} - a_{j_1 j_1 \alpha_2})^{m_{\alpha_2}} \prod_{\beta=1}^{q} (\overline{x}_{j_2} - a_{j_1 j_2 \beta})^{n_\beta} > 0$$

and

$$a_{j_1 j_1 0} \prod_{\alpha_2=1,\alpha_2\neq\alpha_1}^{p} (a_{j_1 j_1 \alpha_1} - a_{j_1 j_1 \alpha_2})^{m_{\alpha_2}} \prod_{\beta=1}^{q} (\overline{x}_{j_2} - a_{j_1 j_2 \beta})^{n_\beta} < 0$$

for $m = 2m_1 + 1$.

The flows at $x_{j_1}^* = a_{j_1 j_1 1}$ and $\overline{x}_{j_2} \neq a_{j_1 j_2 1}$ are the $(2m_1)^{\text{th}}$-order upper-saddle and lower-saddle for

$$a_{j_1 j_1 0} \prod_{\alpha_2=1,\alpha_2\neq\alpha_1}^{p} (a_{j_1 j_1 \alpha_1} - a_{j_1 j_1 \alpha_2})^{m_{\alpha_2}} \prod_{\beta=1}^{q} (\overline{x}_{j_2} - a_{j_1 j_2 \beta})^{n_\beta} > 0$$

and

$$a_{j_1 j_1 0} \prod_{\alpha_2=1,\alpha_2\neq\alpha_1}^{p} (a_{j_1 j_1 \alpha_1} - a_{j_1 j_1 \alpha_2})^{m_{\alpha_2}} \prod_{\beta=1}^{q} (\overline{x}_{j_2} - a_{j_1 j_2 \beta})^{n_\beta} < 0$$

for $m = 2m_1$.

Define

$$\bigcup_{\alpha=1}^{p}\bigcup_{\beta=1}^{q}\underbrace{(a_{j_1j_1\alpha},a_{j_1j_2\beta})}_{([m_\alpha,m_\beta]\text{XX,xF})}$$

$$=\begin{bmatrix}\underbrace{(a_{j_1j_1 1},a_{j_1j_2 q})}_{([m_1,n_q]\text{XX,xF})} & \cdots & \underbrace{(a_{j_1j_1(p-1)},a_{j_1j_2 q})}_{([m_{p-1},n_q]\text{XX,xF})} & \underbrace{(a_{j_1j_1 p},a_{j_1j_2 q})}_{([m_p,n_q]\text{XX,xF})}\\ \underbrace{(a_{j_1j_1 1},a_{j_1j_2(q-1)})}_{([m_1,n_q]\text{XX,xF})} & \cdots & \underbrace{(a_{j_1j_1(p-1)},a_{j_1j_2(q-1)})}_{([m_{p-1},n_q]\text{XX,xF})} & \underbrace{(a_{j_1j_1 p},a_{j_1j_2(q-1)})}_{([m_p,n_{q_{-1}}]\text{XX,xF})}\\ \vdots & \vdots & \vdots & \vdots\\ \underbrace{(a_{j_1j_1 1},a_{j_1j_2 1})}_{([m_1,n_1]\text{XX,xF})} & \cdots & \underbrace{(a_{j_1j_1(p-1)},a_{j_1j_2 1)})}_{([m_{p-1},n_q]\text{XX,xF})} & \underbrace{(a_{j_1j_1 p_{j_1)}},a_{j_1j_2 1})}_{([m_{p_{j_1}},n_1]\text{XX,xF})}\end{bmatrix}.$$

Let $\overline{x}_{j_1} = a_{j_1j_1\alpha_1}$ and $\overline{x}_{j_2} = a_{j_1j_2\beta_1}$. Therefore, from the above discussion, there are four cases: *(iii$_1$)* $m_{\alpha_1} = 2m_{\alpha_1 1}+1$ *and* $n_{\beta_1} = 2n_{\beta_1 1}+1$,*(iii$_2$)* $m_{\alpha_1} = 2m_{\alpha_1 1}$ *and* $n_{\beta_1} = 2n_{\beta_1 1}+1$,*(iii$_3$)* $m_{\alpha_1} = 2m_{\alpha_1 1}+1$ *and* $n_{\beta_1} = 2n_{\beta_1 1}$,*(iii$_4$)* $m_{\alpha_1} = 2m_{\alpha_1 1}$ *and* $n_{\beta_1} = 2n_{\beta_1 1}$.

(iii$_1$) For $m_{\alpha_1} = 2m_{\alpha_1 1}+1$ and $n_{\beta_1} = 2n_{\beta_1 1}+1$, the equilibriums of $(x^*_{j_1}, x^*_{j_2}) = (a_{j_1j_1\alpha_1}, a_{j_1j_2\beta_1})(\alpha_1 \in \{1,2,\ldots,p\};\ \beta_1 \in \{1,2,\ldots,q\})$ for the x_{j_1}-direction has the following properties as in Eqs. (1.42–1.45).

- *For* $a_{j_1j_10}\prod_{\alpha_1=1,\alpha_2\neq\alpha_1}^{p}(a_{j_1j_1\alpha_1}-a_{j_1j_1\alpha_2})^{m_{\alpha_2}}\prod_{\beta_2=1,\beta_2\neq\beta_1}^{q}(a_{j_1j_2\beta_1}-a_{j_1j_2\beta_2})^{n_{\beta_2}}>0$ *and* $a_{j_2j_20}>0$,

$$(a_{j_1j_1\alpha_1}, a_{j_1j_2\beta_1}) = \underbrace{([(2m_{\alpha_1 1}+1),(2n_{\beta_1 1}+1)]\text{DP:UP,pF})}_{[(2m_{\alpha_1 1}+1),(2n_{\beta_1 1}+1)]\text{ - hyperbolic flow }(+)}.$$

  The flow at $(x^*_{j_1}, x^*_{j_2}) = (a_{j_1j_1\alpha_1}, a_{j_1j_2\beta_1})$ is a $([(2m_{\alpha_1 1}+1),(2n_{\beta_1 1}+1)]$DP:UP,pF)-positive hyperbolic flow. The line of $x^*_{j_1} = a_{j_1j_1\alpha_1}$ is a $(2m_{\alpha_1 1}+1)^{\text{th}}$-order positive sink-to-source directrix flow.

- *For* $a_{j_1j_10}\prod_{\alpha_1=1,\alpha_2\neq\alpha_1}^{p}(a_{j_1j_1\alpha_1}-a_{j_1j_1\alpha_2})^{m_{\alpha_2}}\prod_{\beta_2=1,\beta_2\neq\beta_1}^{q}(a_{j_1j_2\beta_1}-a_{j_1j_2\beta_2})^{n_{\beta_2}}<0$ *and* $a_{j_2j_20}>0$,

$$(a_{j_1j_1\alpha_1}, a_{j_1j_2\beta_1}) = \underbrace{([(2m_{\alpha_1 1}+1),(2n_{\beta_1 1}+1)]\text{UP:DP,pF})}_{[(2m_{\alpha_1 1}+1),(2n_{\beta_1 1}+1)]\text{-hyperbolic-secant flow }(+)}.$$

  The flow at $(x^*_{j_1}, x^*_{j_2}) = (a_{j_1j_1\alpha_1}, a_{j_1j_2\beta_1})$ is a $([(2m_{\alpha_1 1}+1),(2n_{\beta_1 1}+1)]$UP : DP, pF)-positive hyperbolic-secant flow. The line of $x^*_{j_1} = a_{j_1j_1\alpha_1}$ is a $(2m_{\alpha_1 1}+1)^{\text{th}}$-order positive source-to-sink directrix flow.

- *For* $a_{j_1j_10}\prod_{\alpha_1=1,\alpha_2\neq\alpha_1}^{p}(a_{j_1j_1\alpha_1}-a_{j_1j_1\alpha_2})^{m_{\alpha_2}}\prod_{\beta_2=1,\beta_2\neq\beta_1}^{q}(a_{j_1j_2\beta_1}-a_{j_1j_2\beta_2})^{n_{\beta_2}}>0$ *and* $a_{j_2j_20}<0$,

$$(a_{j_1j_1\alpha_1},a_{j_1j_2\beta_1})=\underbrace{([(2m_{\alpha_1 1}+1),(2n_{\beta_1 1}+1)]\text{UP:DP,nF})}_{[(2m_{\alpha_1 1}+1),(2n_{\beta_1 1}+1)]\text{-hyperbolic-secant flow }(-)}.$$

The flow at $(x_{j_1}^*,x_{j_2}^*)=(a_{j_1j_1\alpha_1},a_{j_1j_2\beta_1})$ is a $([(2m_{\alpha_1 1}+1),(2n_{\beta_1 1}+1)]$UP : DP, nF)-negative hyperbolic-secant flow. The line of $x_{j_1}^*=a_{j_1j_1\alpha_1}$ is a $(2m_{\alpha_1 1}+1)^{\text{th}}$-order negative sink-to-source directrix flow.

- *For* $a_{j_1j_10}\prod_{\alpha_1=1,\alpha_2\neq\alpha_1}^{p}(a_{j_1j_1\alpha_1}-a_{j_1j_1\alpha_2})^{m_{\alpha_2}}\prod_{\beta_2=1,\beta_2\neq\beta_1}^{q}(a_{j_1j_2\beta_1}-a_{j_1j_2\beta_2})^{n_{\beta_2}}<0$ *and* $a_{j_2j_20}<0$,

$$(a_{j_1j_1\alpha_1},a_{j_1j_2\beta_1})=\underbrace{([(2m_{\alpha_1 1}+1),(2n_{\beta_1 1}+1)]\text{DP:UP,nF})}_{[(2m_{\alpha_1 1}+1),(2n_{\beta_1 1}+1)]\text{-hyperbolic flow }(-)}.$$

The flow at $(x_{j_1}^*,x_{j_2}^*)=(a_{j_1j_1\alpha_1},a_{j_1j_2\beta_1})$ is a $([(2m_{\alpha_1 1}+1),(2n_{\beta_1 1}+1)]$DP:UP,nF)-negative hyperbolic flow. The line of $x_{j_1}^*=a_{j_1j_1\alpha_1}$ is a $(2m_{\alpha_1 1}+1)^{\text{th}}$-order negative source-to-sink directrix flow.

(iii$_2$) For $m_{\alpha_1}=2m_{\alpha_1 1}$ *and* $n_{\beta_1}=2n_{\beta_1 1}+1$, *the equilibriums of* $(x_{j_1}^*,x_{j_2}^*)=(a_{j_1j_1\alpha_1},a_{j_1j_2\beta_1})(\alpha_1\in\{1,2,\ldots,p\};\beta_1\in\{1,2,\ldots,q\})$ *for the* x_{j_1}*-direction has the following properties as Eqs. (*1.46–1.49*).*

- *For* $a_{j_1j_10}\prod_{\alpha_1=1,\alpha_2\neq\alpha_1}^{p}(a_{j_1j_1\alpha_1}-a_{j_1j_1\alpha_2})^{m_{\alpha_2}}\prod_{\beta_2=1,\beta_2\neq\beta_1}^{q}(a_{j_1j_2\beta_1}-a_{j_1j_2\beta_2})^{n_{\beta_2}}>0$ *and* $a_{j_2j_20}>0$,

$$(a_{j_1j_1\alpha_1},a_{j_1j_2\beta_1})=\underbrace{([(2m_{\alpha_1 1}),(2n_{\beta_1 1}+1)]\text{UP:UP,pF})}_{[(2m_{\alpha_1 1}),(2n_{\beta_1 1}+1)]\text{-hyperbolic-secant-to-hyperbolic flow }(+)}.$$

The flow at $(x_{j_1}^*,x_{j_2}^*)=(a_{j_1j_1\alpha_1},a_{j_1j_2\beta_1})$ *is a* $([(2m_{\alpha_1 1}),(2n_{\beta_1 1}+1)]$UP:UP,pF)-positive hyperbolic-secant-to-hyperbolic flow. The line of $x_{j_1}^*=a_{j_1j_1\alpha_1}$ is a $(2m_{\alpha_1 1})^{\text{th}}$-order positive lower-to-upper saddle directrix flow.

- *For* $a_{j_1j_10}\prod_{\alpha_1=1,\alpha_2\neq\alpha_1}^{p}(a_{j_1j_1\alpha_1}-a_{j_1j_1\alpha_2})^{m_{\alpha_2}}\prod_{\beta_2=1,\beta_2\neq\beta_1}^{q}(a_{j_1j_2\beta_1}-a_{j_1j_2\beta_2})^{n_{\beta_2}}<0$ *and* $a_{j_2j_20}>0$,

$$(a_{j_1j_1\alpha_1},a_{j_1j_2\beta_1})=\underbrace{([(2m_{\alpha_1 1}),(2n_{\beta_1 1}+1)]\text{DP:DP,pF})}_{[(2m_{\alpha_1 1}),(2n_{\beta_1 1}+1)]\text{-hyperbolic-to-hyperbolic-secant flow }(+)}.$$

The flow at $(x_{j_1}^*, x_{j_2}^*) = (a_{j_1j_1\alpha_1}, a_{j_1j_2\beta_1})$ is a ([$(2m_{\alpha_1 1}), (2n_{\beta_1 1}+1)$]UP:UP,pF)-positive hyperbolic-to-hyperbolic-secant flow. The line of $x_{j_1}^* = a_{j_1j_1\alpha_1}$ is a $(2m_{\alpha_1 1})^{\text{th}}$-order positive upper-to-lower saddle directrix flow.

- *For* $a_{j_1j_10}\prod_{\alpha_1=1,\alpha_2\neq\alpha_1}^{p}(a_{j_1j_1\alpha_1}-a_{j_1j_1\alpha_2})^{m_{\alpha_2}}\prod_{\beta_2=1,\beta_2\neq\beta_1}^{q}(a_{j_1j_2\beta_1}-a_{j_1j_2\beta_2})^{n_{\beta_2}}>0$ *and* $a_{j_2j_20}<0$,

$$(a_{j_1j_1\alpha_1}, a_{j_1j_2\beta_1}) = \underbrace{([(2m_{\alpha_1 1}), (2n_{\beta_1 1}+1)]\text{DP:DP,nF})}_{[(2m_{\alpha_1 1}),(2n_{\beta_1 1}+1)]\text{-hyperbolic-to-hyperbolic-secant flow }(-)}.$$

The flow at $(x_{j_1}^*, x_{j_2}^*) = (a_{j_1j_1\alpha_1}, a_{j_1j_2\beta_1})$ is a ([$(2m_{\alpha_1 1}), (2n_{\beta_1 1}+1)$]DP:DP,nF)-negative hyperbolic-to-hyperbolic-secant flow. The line of $x_{j_1}^* = a_{j_1j_1\alpha_1}$ is a $(2m_{\alpha_1 1})^{\text{th}}$-order negative lower-to-upper saddle directrix flow.

- *For* $a_{j_1j_10}\prod_{\alpha_1=1,\alpha_2\neq\alpha_1}^{p}(a_{j_1j_1\alpha_1}-a_{j_1j_1\alpha_2})^{m_{\alpha_2}}\prod_{\beta_2=1,\beta_2\neq\beta_1}^{q}(a_{j_1j_2\beta_1}-a_{j_1j_2\beta_2})^{n_{\beta_2}}<0$ and $a_{j_2j_20}<0$,

$$(a_{j_1j_1\alpha_1}, a_{j_1j_2\beta_1}) = \underbrace{([(2m_{\alpha_1 1}), (2n_{\beta_1 1}+1)]\text{UP:UP,nF})}_{[(2m_{\alpha_1 1}),(2n_{\beta_1 1}+1)]\text{-hyperbolic-secant-to-hyperbolic flow }(-)}.$$

The flow at $(x_{j_1}^*, x_{j_2}^*) = (a_{j_1j_1\alpha_1}, a_{j_1j_2\beta_1})$ is a ([$(2m_{\alpha_1 1}), (2n_{\beta_1 1}+1)$]UP:UP,nF)-negative hyperbolic-secant-to-hyperbolic flow. The line of $x_{j_1}^* = a_{j_1j_1\alpha_1}$ is a $(2m_{\alpha_1 1})^{\text{th}}$-order negative upper-to-lower saddle directrix flow.

(iii₃) For $m_{\alpha_1} = 2m_{\alpha_1 1}+1$ *and* $n_{\beta_1} = 2n_{\beta_1 1}$, *the equilibriums of* $(x_{j_1}^*, x_{j_2}^*) = (a_{j_1j_1\alpha_1}, a_{j_1j_2\beta_1})(\alpha_1 \in \{1,2,\ldots,p\}; \beta_1 \in \{1,2,\ldots,q\})$ *for the* x_{j_1}*-direction has the following properties as in Eqs. (*1.50–1.53*).*

- *For* $a_{j_1j_10}\prod_{\alpha_1=1,\alpha_2\neq\alpha_1}^{p}(a_{j_1j_1\alpha_1}-a_{j_1j_1\alpha_2})^{m_{\alpha_2}}\prod_{\beta_2=1,\beta_2\neq\beta_1}^{q}(a_{j_1j_2\beta_1}-a_{j_1j_2\beta_2})^{n_{\beta_2}}>0$ *and* $a_{j_2j_20}>0$,

$$(a_{j_1j_1\alpha_1}, a_{j_1j_2\beta_1}) = \underbrace{([(2m_{\alpha_1 1}+1), (2n_{\beta_1 1})]_{\text{DI:II}}\text{SO,pF})}_{[(2m_{\alpha_1 1}+1),(2n_{\beta_1 1})]\text{-increasing-to-decreasing inflection source flow }(+)}.$$

The flow at $(x_{j_1}^*, x_{j_2}^*) = (a_{j_1j_1\alpha_1}, a_{j_1j_2\beta_1})$ is a ([$(2m_{\alpha_1 1}+1), (2n_{\beta_1 1})$]$_{\text{DI:II}}$SO,pF)-positive decreasing-to-increasing inflection source flow. The line of $x_{j_1}^* = a_{j_1j_1\alpha_1}$ is a $(2m_{\alpha_1 1}+1)^{\text{th}}$-order positive source directrix flow.

- *For* $a_{j_1j_10}\prod_{\alpha_1=1,\alpha_2\neq\alpha_1}^{p}(a_{j_1j_1\alpha_1}-a_{j_1j_1\alpha_2})^{m_{\alpha_2}}\prod_{\beta_2=1,\beta_2\neq\beta_1}^{q}(a_{j_1j_2\beta_1}-a_{j_1j_2\beta_2})^{n_{\beta_2}}<0$ *and* $a_{j_2j_20}>0$,

$$(a_{j_1j_1\alpha_1},a_{j_1j_2\beta_1})=\underbrace{([(2m_{\alpha_1 1}+1),(2n_{\beta_1 1})]_{\text{II:DI}}\text{SI,pF})}_{[(2m_{\alpha_1 1}+1),(2n_{\beta_1 1})]\text{-decreasing-to-increasing inflection sink flow }(+)}.$$

The flow at $(x_{j_1}^*,x_{j_2}^*)=(a_{j_1j_1\alpha_1},a_{j_1j_2\beta_1})$ is a $([(2m_{\alpha_1 1}+1),(2n_{\beta_1 1})]_{\text{II:DI}}\text{SI,pF})$-positive increasing-to-decreasing inflection sink flow. The line of $x_{j_1}^*=a_{j_1j_1\alpha_1}$ is a $(2m_{\alpha_1 1}+1)^{\text{th}}$-order positive sink directrix flow.

- *For* $a_{j_1j_10}\prod_{\alpha_1=1,\alpha_2\neq\alpha_1}^{p}(a_{j_1j_1\alpha_1}-a_{j_1j_1\alpha_2})^{m_{\alpha_2}}\prod_{\beta_2=1,\beta_2\neq\beta_1}^{q}(a_{j_1j_2\beta_1}-a_{j_1j_2\beta_2})^{n_{\beta_2}}>0$ *and* $a_{j_2j_20}<0$,

$$(a_{j_1j_1\alpha_1},a_{j_1j_2\beta_1})=\underbrace{([(2m_{\alpha_1 1}+1),(2n_{\beta_1 1})]_{\text{II:DI}}\text{SO,nF})}_{[(2m_{\alpha_1 1}+1),(2n_{\beta_1 1})]\text{-decreasing-to-increasing inflection source flow }(-)}.$$

The flow at $(x_{j_1}^*,x_{j_2}^*)=(a_{j_1j_1\alpha_1},a_{j_1j_2\beta_1})$ is a $([(2m_{\alpha_1 1}+1),(2n_{\beta_1 1})]_{\text{II:DI}}\text{SO,nF})$-negative increasing-to-decreasing inflection source flow. The line of $x_{j_1}^*=a_{j_1j_1\alpha_1}$ is a $(2m_{\alpha_1 1}+1)^{\text{th}}$-order negative source directrix flow.

- *For* $a_{j_1j_10}\prod_{\alpha_1=1,\alpha_2\neq\alpha_1}^{p}(a_{j_1j_1\alpha_1}-a_{j_1j_1\alpha_2})^{m_{\alpha_2}}\prod_{\beta_2=1,\beta_2\neq\beta_1}^{q}(a_{j_1j_2\beta_1}-a_{j_1j_2\beta_2})^{n_{\beta_2}}<0$ *and* $a_{j_2j_20}<0$,

$$(a_{j_1j_1\alpha_1},a_{j_1j_2\beta_1})=\underbrace{([(2m_{\alpha_1 1}+1),(2n_{\beta_1 1})]_{\text{DI:II}}\text{SI,nF})}_{[(2m_{\alpha_1 1}+1),(2n_{\beta_1 1})]\text{-increasing-to-decreasing inflection sink flow }(-)}.$$

The flow at $(x_{j_1}^*,x_{j_2}^*)=(a_{j_1j_1\alpha_1},a_{j_1j_2\beta_1})$ is a $([(2m_{\alpha_1 1}+1),(2n_{\beta_1 1})]_{\text{DI:II}}\text{SI,nF})$-negative decreasing-to-increasing inflection sink flow. The line of $x_{j_1}^*=a_{j_1j_1\alpha_1}$ is a $(2m_{\alpha_1 1}+1)^{\text{th}}$-order negative sink directrix flow.

(iii$_4$) For $m_{\alpha_1}=2m_{\alpha_1 1}$ *and* $n_{\beta_1}=2n_{\beta_1 1}$, *the equilibriums of* $(x_{j_1}^*,x_{j_2}^*)=(a_{j_1j_1\alpha_1},a_{j_1j_2\beta_1})(\alpha_1\in\{1,2,\ldots,p\};\beta_1\in\{1,2,\ldots,q\})$ *for the* x_{j_1}*-direction has the following properties as in Eqs. (*1.54–1.57*).*

- *For* $a_{j_1j_10}\prod_{\alpha_1=1,\alpha_2\neq\alpha_1}^{p}(a_{j_1j_1\alpha_1}-a_{j_1j_1\alpha_2})^{m_{\alpha_2}}\prod_{\beta_2=1,\beta_2\neq\beta_1}^{q}(a_{j_1j_2\beta_1}-a_{j_1j_2\beta_2})^{n_{\beta_2}}>0$ and $a_{j_2j_20}>0$,

$$(a_{j_1j_1\alpha_1},a_{j_1j_2\beta_1})=\underbrace{([(2m_{\alpha_1 1}),(2n_{\beta_1 1})]_{\text{II:II}}\text{US,pF})}_{[(2m_{\alpha_1 1}),(2n_{\beta_1 1})]\text{-increasing-inflection upper-saddle flow }(+)}.$$

The flow at $(x_{j_1}^*, x_{j_2}^*) = (a_{j_1j_1\alpha_1}, a_{j_1j_2\beta_1})$ is a $([(2m_{\alpha_1}1), (2n_{\beta_1}1)]_{\text{II:II}}\text{US,pF})$-positive increasing-inflection upper-saddle flow. The line of $x_{j_1}^* = a_{j_1j_1\alpha_1}$ is a $(2m_{\alpha_1}1)^{\text{th}}$-order positive upper-saddle directrix flow.

- *For* $a_{j_1j_10}\prod_{\alpha_1=1,\alpha_2\neq\alpha_1}^{p}(a_{j_1j_1\alpha_1}-a_{j_1j_1\alpha_2})^{m_{\alpha_2}}\prod_{\beta_2=1,\beta_2\neq\beta_1}^{q}(a_{j_1j_2\beta_1}-a_{j_1j_2\beta_2})^{n_{\beta_2}} < 0$ *and* $a_{j_2j_20} > 0$,

$$(a_{j_1j_1\alpha_1}, a_{j_1j_2\beta_1}) = \underbrace{([(2m_{\alpha_1}1), (2n_{\beta_1}1)]_{\text{DI:DI}}\text{LS,pF})}_{[(2m_{\alpha_1}1),(2n_{\beta_1}1)]\text{-decreasing-inflection lower-saddle flow }(+)}.$$

The flow at $(x_{j_1}^*, x_{j_2}^*) = (a_{j_1j_1\alpha_1}, a_{j_1j_2\beta_1})$ is a $([(2m_{\alpha_1}1), (2n_{\beta_1}1)]_{\text{DI:DI}}\text{LS,pF})$-positive decreasing-inflection lower-saddle flow. The line of $x_{j_1}^* = a_{j_1j_1\alpha_1}$ is a $(2m_{\alpha_1}1)^{\text{th}}$-order positive lower-saddle directrix flow.

- *For* $a_{j_1j_10}\prod_{\alpha_1=1,\alpha_2\neq\alpha_1}^{p}(a_{j_1j_1\alpha_1}-a_{j_1j_1\alpha_2})^{m_{\alpha_2}}\prod_{\beta_2=1,\beta_2\neq\beta_1}^{q}(a_{j_1j_2\beta_1}-a_{j_1j_2\beta_2})^{n_{\beta_2}} > 0$ *and* $a_{j_2j_20} < 0$,

$$(a_{j_1j_1\alpha_1}, a_{j_1j_2\beta_1}) = \underbrace{([(2m_{\alpha_1}1), (2n_{\beta_1}1)]_{\text{DI:DI}}\text{US,nF})}_{[(2m_{\alpha_1}1),(2n_{\beta_1}1)]\text{-decreasing-inflection upper-saddle flow }(-)}.$$

The flow at $(x_{j_1}^*, x_{j_2}^*) = (a_{j_1j_1\alpha_1}, a_{j_1j_2\beta_1})$ is a $([(2m_{\alpha_1}1), (2n_{\beta_1}1)]_{\text{DI:DI}}\text{US,nF})$-negative decreasing-inflection upper-saddle flow. The line of $x_{j_1}^* = a_{j_1j_1\alpha_1}$ is a $(2m_{\alpha_1}1)^{\text{th}}$-order negative upper-saddle directrix flow.

- *For* $a_{j_1j_10}\prod_{\alpha_1=1,\alpha_2\neq\alpha_1}^{p}(a_{j_1j_1\alpha_1}-a_{j_1j_1\alpha_2})^{m_{\alpha_2}}\prod_{\beta_2=1,\beta_2\neq\beta_1}^{q}(a_{j_1j_2\beta_1}-a_{j_1j_2\beta_2})^{n_{\beta_2}} < 0$ *and* $a_{j_2j_20} < 0$,

$$(a_{j_1j_1\alpha_1}, a_{j_1j_2\beta_1}) = \underbrace{([(2m_{\alpha_1}1), (2n_{\beta_1}1)]_{\text{II:II}}\text{LS,nF})}_{[(2m_{\alpha_1}1),(2n_{\beta_1}1)]\text{-increasing-inflection lower-saddle flow }(-)}.$$

The flow at $(x_{j_1}^*, x_{j_2}^*) = (a_{j_1j_1\alpha_1}, a_{j_1j_2\beta_1})$ is a $([(2m_{\alpha_1}1), (2n_{\beta_1}1)]_{\text{II:II}}\text{LS,nF})$-negative increasing-inflection lower-saddle flow. The line of $x_{j_1}^* = a_{j_1j_1\alpha_1}$ is a $(2m_{\alpha_1}1)^{\text{th}}$-order negative lower-saddle directrix flow.

2.4 Hyperbolic and Hyperbolic-Secant Networks

(iv) Consider

$$\dot{x}_{j_2} = a_{j_2 j_2 0}$$
$$\dot{x}_{j_1} = a_{j_1 j_1 0} \prod_{\alpha=1}^{m} (x_{j_1} - a_{j_1 j_1 \alpha}) \prod_{\beta=1}^{q} (x_{j_2} - a_{j_1 j_2 \beta}).$$

The foregoing equation becomes

$$\frac{dx_{j_1}}{dx_{j_2}} = \frac{a_{j_1 j_1 0}}{a_{j_2 j_2 0}} \prod_{\alpha=1}^{m} (x_{j_1} - a_{j_1 j_1 \alpha}) \prod_{\beta=1}^{q} (x_{j_2} - a_{j_1 j_2 \beta}),$$

The deformation gives

$$\frac{a_{j_2 j_2 0} dx_{j_1}}{\prod\limits_{\alpha=1}^{m} (x_{j_1} - a_{j_1 j_1 \alpha})} = a_{j_1 j_1 0} \prod_{\beta=1}^{n} (x_{j_2} - a_{j_1 j_2 \beta}) dx_{j_2},$$

or

$$a_{j_2 j_2 0} \sum_{\alpha_1=1}^{m} \frac{dx_{j_1}}{(x_{j_1} - a_{j_1 j_1 \alpha_1}) \prod_{\alpha_2=1, \alpha_2 \neq \alpha_1}^{m} (a_{j_1 j_1 \alpha_1} - a_{j_1 j_1 \alpha_2})}$$
$$= a_{j_1 j_1 0} (x_{j_2} - a_{j_1 j_2 \beta_1}) \prod_{\beta_2=1, \beta_2 \neq \beta_2}^{m} (x_{j_2} - a_{j_1 j_2 \beta_1} + a_{j_1 j_2 \beta_1} - a_{j_1 j_2 \beta}) dx_{j_2}$$

and

$$a_{j_2 j_2 0} \sum_{\alpha_1=1}^{m} \frac{dx_{j_1}}{(x_{j_1} - a_{j_1 j_1 \alpha_1}) \prod_{\alpha_2=1, \alpha_2 \neq \alpha_1}^{m} (a_{j_1 j_1 \alpha_1} - a_{j_1 j_1 \alpha_2})}$$
$$= a_{j_1 j_1 0} [(x_{j_2} - a_{j_1 j_2 \beta_1})^n + \sum_{\rho=1}^{n-1} b_{j_1 j_2 \rho} (x_{j_2} - a_{j_1 j_2 \beta_1})^{n-\rho}] dx_{j_2},$$

where

$$b_{j_1j_21} = \sum_{\beta_2=1,\beta_2\neq\beta_1}^{m} (a_{j_1j_2\beta_1} - a_{j_1j_2\beta_2}),$$

$$b_{j_1j_22} = \sum_{\substack{\beta_2,\beta_3=1;\beta_2,\beta_3\neq\beta_1\\(\beta_2<\beta_3)}}^{m} \prod_{r=2}^{3} (a_{j_1j_2\beta_1} - a_{j_1j_2\beta_r}), \ldots,$$

$$b_{j_1j_2k} = \sum_{\substack{\beta_2,\beta_3,\ldots,\beta_{k+1}=1;\\ \beta_2,\beta_3,\ldots,\beta_{k+1}\neq\beta_1\\(\beta_2<\beta_3<\ldots<\beta_{k+1})}}^{m} \prod_{r=2}^{k+1} (a_{j_1j_2\beta_1} - a_{j_1j_2\beta_r}), \ldots,$$

$$b_{j_1j_2(m-1)} = \prod_{\beta_1=1,\beta_2\neq s_1}^{m} (a_{j_1j_2\beta_1} - a_{j_1j_2\beta_2}).$$

with an initial condition $(x_{j_1 0}, x_{j_2 0})$ art time $t = t_0$, integration of the foregoing equation gives the first integral manifold as

$$a_{j_2j_20} \sum_{\alpha_1=1}^{m} \frac{1}{\prod_{\alpha_2=1,\alpha_2\neq\alpha_1}^{m} (a_{j_1j_1\alpha_1} - a_{j_1j_1\alpha_2})} \ln \frac{|x_{j_1} - a_{j_1j_1\alpha_1}|}{|x_{j_10} - a_{j_1j_1\alpha_1}|}$$
$$= a_{j_1j_10}\Big\{\frac{1}{n+1}\big[(x_{j_2} - a_{j_1j_2\beta_1})^{n+1} - (x_{j_20} - a_{j_1j_2\beta_1})^{n+1}\big]$$
$$+ \sum_{k=1}^{n-1} \frac{1}{n-k+1} b_{j_1j_2k}\big[(x_{j_2} - a_{j_1j_2\beta_1})^{n-k+1} - (x_{j_20} - a_{j_1j_2\beta_1})^{n-k+1}\big]\Big\}.$$

At the equilibrium of $x^*_{j_2} = a_{j_1j_2\beta_1}$ $(\beta_1, \beta_2 \in \{1, 2, \ldots, n\}, \beta_1 \neq \beta_2)$ and $\overline{x}_{j_1} \neq a_{j_1j_1\alpha_1}$ $(\alpha = 1, 2, \ldots, m)$, in phase space,

$$\frac{dx_{j_1}}{dx_{j_2}}\Big|_{x^*_{j_2}=a_{j_1j_2\beta_1}} = \frac{a_{j_1j_10}}{a_{j_2j_20}} \prod_{\alpha=1}^{m} (\overline{x}_{j_1} - a_{j_1j_1\alpha})$$
$$\times (x_{j_2} - a_{j_1j_2\beta_1}) \prod_{\beta_2=1,\beta_2\neq\beta_1}^{n} (x_{j_2} - a_{j_1j_2\beta_2})\Big|_{x^*_{j_2}=a_{j_1j_2\beta_1}}$$
$$= 0.$$

If

$$\frac{d^2x_{j_1}}{dx^2_{j_2}}\Big|_{x^*_{j_2}=a_{j_1j_2\beta_1}} = \frac{a_{j_1j_10}}{a_{j_2j_20}} \prod_{\alpha=1}^{m} (\overline{x}_{j_1} - a_{j_1j_1\alpha}) \prod_{\beta_2=1,\beta_2\neq\beta_1}^{q} (a_{j_1j_2\beta_1} - a_{j_1j_2\beta_2})$$
$$> 0,$$

the up-parabola flow exists. If

$$\left.\frac{d^{n_{\beta_1}+1}x_{j_1}}{dx_{j_2}^{n_{\beta_1}+1}}\right|_{x_{j_2}^*=a_{j_1j_2\beta_1}} = \frac{a_{j_1j_10}}{a_{j_2j_20}}\prod_{\alpha=1}^{m}(\overline{x}_{j_1}-a_{j_1j_1\alpha})\prod_{\beta_2=1,\beta_2\neq\beta_2}^{n}(a_{j_1j_2\beta_1}-a_{j_1j_2\beta_2})$$
$$< 0,$$

the down-parabola flow exists.

At the equilibrium of $x_{j_1}^* = a_{j_1j_1\alpha_1}$ ($\alpha_1 \in \{1, 2, \ldots, m\}$) and $\overline{x}_{j_2} \neq a_{j_1j_2\beta}$ ($\beta \in \{1, 2, \ldots, n\}$) the variational equation is

$$\dot{x}_{j_2} = a_{j_2j_20},$$
$$\Delta\dot{x}_{j_1} = a_{j_1j_10}\prod_{\alpha_2=1,\alpha_2\neq\alpha_1}^{m}(a_{j_1j_1\alpha_1}-a_{j_1j_1\alpha_2})\prod_{\beta=1}^{n}(\overline{x}_{j_2}-a_{j_1j_2\beta})\Delta x_{j_1}.$$

The flows at $x_{j_1}^* = a_{j_1j_1\alpha_1}$ and $\overline{x}_{j_2} \neq a_{j_1j_2\beta}$ are source and sink for

$$a_{j_1j_10}\prod_{\alpha_2=1,\alpha_2\neq\alpha_1}^{m}(a_{j_1j_1\alpha_1}-a_{j_1j_1\alpha_2})\prod_{\beta=1}^{n}(\overline{x}_{j_2}-a_{j_1j_2\beta}) > 0,$$

and

$$a_{j_1j_10}\prod_{\alpha_2=1,\alpha_2\neq\alpha_1}^{m}(a_{j_1j_1\alpha_1}-a_{j_1j_1\alpha_2})\prod_{\beta=1}^{n}(\overline{x}_{j_2}-a_{j_1j_2\beta}) < 0.$$

(iv_1) The equilibrium of $(x_{j_1}^*, x_{j_2}^*) = (a_{j_1j_1\alpha_1}, a_{j_1j_2\beta_1})$ ($\alpha_1 \in \{1, 2, \ldots, m\}, \beta_1 \in \{1, 2, \ldots, n\}$) for the x_{j_1}-direction has the following properties as in Eqs. (1.61–1.64).

- For $a_{j_1j_10}\prod_{\alpha_2=1,\alpha_2\neq\alpha_1}^{m}(a_{j_1j_1\alpha_1}-a_{j_1j_1\alpha_2})\prod_{\beta_2=1,\beta_2\neq\beta_1}^{n}(a_{j_1j_2\beta_2}-a_{j_1j_2\beta_2}) > 0$ and $a_{j_2j_20} > 0$

$$(a_{j_1j_1\alpha_1}, a_{j_1j_2\beta_1}) = \underbrace{(\text{DP:UP,pF})}_{\text{hyperbolic flow }(+)}.$$

- For $a_{j_1j_10}\prod_{\alpha_2=1,\alpha_2\neq\alpha_1}^{m}(a_{j_1j_1\alpha_1}-a_{j_1j_1\alpha_2})\prod_{\beta_2=1,\beta_2\neq\beta_1}^{n}(a_{j_1j_2\beta_2}-a_{j_1j_2\beta_2}) < 0$ and $a_{j_2j_20} > 0$

$$(a_{j_1j_1\alpha_1}, a_{j_1j_2\beta_1}) = \underbrace{(\text{UP:DP,pF})}_{\text{hyperbolic-secant flow }(+)}.$$

- For $a_{j_1j_10}\prod_{\alpha_2=1,\alpha_2\neq\alpha_1}^{m}(a_{j_1j_1\alpha_1}-a_{j_1j_1\alpha_2})\prod_{\beta_2=1,\beta_2\neq\beta_1}^{n}(a_{j_1j_2\beta_2}-a_{j_1j_2\beta_2}) > 0$ and $a_{j_2j_20} < 0$

$$(a_{j_1j_1\alpha_1}, a_{j_1j_2\beta_1}) = \underbrace{(\text{UP:DP,nF})}_{\text{hyperbolic-secant flow }(-)}.$$

- For $a_{j_1j_10}\prod_{\alpha_2=1,\alpha_2\neq\alpha_1}^{m}(a_{j_1j_1\alpha_1}-a_{j_1j_1\alpha_2})\prod_{\beta_2=1,\beta_2\neq\beta_1}^{n}(a_{j_1j_2\beta_2}-a_{j_1j_2\beta_2}) < 0$ and $a_{j_2j_20} < 0$

$$(a_{j_1j_1\alpha_1}, a_{j_1j_2\beta_1}) = \underbrace{\text{(DP:UP,nF)}}_{\text{hyperbolic flow }(-)}.$$

(iv_2)There are four cases: *(iii_{2a}) $m = 2m_1 + 1$ and $n = 2n_1 + 1$,(iii_{2b}) $m = 2m_1$ and $n = 2n_1 + 1$, (iii_{2c}) $m = 2m_1 + 1$ and $n = 2n_1$, (iii_{2d}) $m = 2m_1$ and $n = 2n_1$.*
Define

$$\bigcup_{\alpha=1}^{m}\bigcup_{\beta=1}^{n}\underbrace{(a_{j_1j_1\alpha}, a_{j_1j_2\beta})}_{\text{XX}}$$
$$= \begin{bmatrix} (a_{j_1j_11}, a_{j_1j_2n}) & \cdots & (a_{j_1j_1(m-1)}, a_{j_1j_2n}) & (a_{j_1j_1m}, a_{j_1j_2n}) \\ (a_{j_1j_11}, a_{j_1j_2(n-1)}) & \cdots & (a_{j_1j_1(m-1)}, a_{j_1j_2(n-1)}) & (a_{j_1j_1m}, a_{j_1j_2(n-1)}) \\ \vdots & \vdots & \vdots & \vdots \\ (a_{j_1j_11}, a_{j_1j_21}) & \cdots & (a_{j_1j_1(m-1)}, a_{j_1j_21}) & (a_{j_1j_1m}, a_{j_1j_21}) \end{bmatrix}_{n\times m}.$$

(iv_{2a}) For $m = 2m_1+1$ and $n = 2n_1+1$, the equilibriums of $(x_{j_1}^*, x_{j_2}^*) = (a_{j_1j_1\alpha}, a_{j_1j_2\beta})$ ($\alpha \in \{1, 2, \ldots, 2m_1 + 1\}$,$\beta \in \{1, 2, \ldots, 2n_1 + 1\}$) for the x_{j_1}-direction are a hyperbolic and hyperbolic-secant flow network as in Eqs. (1.66–1.69*).*

- *For $a_{j_1j_10} > 0$ and $a_{j_2j_20} > 0$,*

$$\bigcup_{\alpha=1}^{2m_1+1}\bigcup_{\beta=1}^{2n_1+1}\underbrace{(a_{j_1j_1\alpha}, a_{j_1j_2\beta})}_{\text{(XX, pF)}}$$
$$= \begin{bmatrix} \underbrace{\text{(DP:UP,pF)}}_{\text{hyperbolic flow }(+)} & \cdots & \underbrace{\text{(UP:DP,pF)}}_{\text{hyperbolic-secant flow }(+)} & \underbrace{\text{(DP:UP,pF)}}_{\text{hyperbolic flow }(+)} \\ \underbrace{\text{(UP:DP,pF)}}_{\text{hyperbolic-secant flow }(+)} & \cdots & \underbrace{\text{(DP:UP,pF)}}_{\text{hyperbolic flow }(+)} & \underbrace{\text{(UP:DP,pF)}}_{\text{hyperbolic-secant flow }(+)} \\ \vdots & \vdots & \vdots & \vdots \\ \underbrace{\text{(DP:UP,pF)}}_{\text{hyperbolic flow }(+)} & \cdots & \underbrace{\text{(UP:DP,pF)}}_{\text{hyperbolic-secant flow }(+)} & \underbrace{\text{(DP:UP,pF)}}_{\text{hyperbolic flow }(+)} \end{bmatrix}_{(2n_1+1)\times(2m_1+1)}.$$

- *For* $a_{j_1j_10} < 0$ *and* $a_{j_2j_20} > 0$,

$$\bigcup_{\alpha=1}^{2m_1+1}\bigcup_{\beta=1}^{2n_1+1}\underbrace{(a_{j_1j_1\alpha},a_{j_1j_2\beta})}_{\text{(XX, pF)}} = \begin{bmatrix} \underbrace{\text{(UP:DP,pF)}}_{\text{hyperbolic-secant flow (+)}} & \cdots & \underbrace{\text{(DP:UP,pF)}}_{\text{hyperbolic flow (+)}} & \underbrace{\text{(UP:DP,pF)}}_{\text{hyperbolic-secant flow (+)}} \\ \underbrace{\text{(DP:UP,pF)}}_{\text{hyperbolic flow (+)}} & \cdots & \underbrace{\text{(UP:DP,pF)}}_{\text{hyperbolic-secant flow (+)}} & \underbrace{\text{(DP:UP,pF)}}_{\text{hyperbolic flow (+)}} \\ \vdots & \vdots & \vdots & \vdots \\ \underbrace{\text{(UP:DP,pF)}}_{\text{hyperbolic-secant flow (+)}} & \cdots & \underbrace{\text{(DP:UP,pF)}}_{\text{hyperbolic flow (+)}} & \underbrace{\text{(UP:DP,pF)}}_{\text{hyperbolic-secant flow (+)}} \end{bmatrix}_{(2n_1+1)\times(2m_1+1)}.$$

- *For* $a_{j_1j_10} > 0$ *and* $a_{j_2j_20} < 0$,

$$\bigcup_{\alpha=1}^{2m_1+1}\bigcup_{\beta=1}^{2n_1+1}\underbrace{(a_{j_1j_1\alpha},a_{j_1j_2\beta})}_{\text{(XX, nF)}} = \begin{bmatrix} \underbrace{\text{(UP:DP,nF)}}_{\text{hyperbolic-secant flow (−)}} & \cdots & \underbrace{\text{(DP:UP,nF)}}_{\text{hyperbolic flow (−)}} & \underbrace{\text{(UP:DP,nF)}}_{\text{hyperbolic-secant flow (−)}} \\ \underbrace{\text{(DP:UP,nF)}}_{\text{hyperbolic flow (−)}} & \cdots & \underbrace{\text{(UP:DP,nF)}}_{\text{hyperbolic-secant flow (−)}} & \underbrace{\text{(DP:UP,nF)}}_{\text{hyperbolic flow (−)}} \\ \vdots & \vdots & \vdots & \vdots \\ \underbrace{\text{(UP:DP,nF)}}_{\text{hyperbolic-secant flow (−)}} & \cdots & \underbrace{\text{(DP:UP,nF)}}_{\text{hyperbolic flow (−)}} & \underbrace{\text{(UP:DP,nF)}}_{\text{hyperbolic-secant flow (−)}} \end{bmatrix}_{(2n_1+1)\times(2m_1+1)}.$$

- *For* $a_{j_1j_10} < 0$ *and* $a_{j_2j_20} < 0$,

$$\bigcup_{\alpha=1}^{2m_1+1}\bigcup_{\beta=1}^{2n_1+1}\underbrace{(a_{j_1j_1\alpha},a_{j_1j_2\beta})}_{\text{(XX, nF)}} = \begin{bmatrix} \underbrace{\text{(DP:UP,nF)}}_{\text{hyperbolic flow (−)}} & \cdots & \underbrace{\text{(UP:DP,nF)}}_{\text{hyperbolic-secant flow (−)}} & \underbrace{\text{(DP:UP,nF)}}_{\text{hyperbolic flow (−)}} \\ \underbrace{\text{(UP:DP,nF)}}_{\text{hyperbolic-secant flow (−)}} & \cdots & \underbrace{\text{(DP:UP,nF)}}_{\text{hyperbolic flow (−)}} & \underbrace{\text{(UP:DP,nF)}}_{\text{hyperbolic-secant flow (−)}} \\ \vdots & \vdots & \vdots & \vdots \\ \underbrace{\text{(DP:UP,nF)}}_{\text{hyperbolic flow (−)}} & \cdots & \underbrace{\text{(UP:DP,nF)}}_{\text{hyperbolic-secant flow (−)}} & \underbrace{\text{(DP:UP,nF)}}_{\text{hyperbolic flow (−)}} \end{bmatrix}_{(2n_1+1)\times(2m_1+1)}.$$

The lines of $x_{j_1}^* = a_{j_1j_1\alpha}$ $(\alpha \in \{1, 2, \ldots, 2m_1 + 1\})$ *are directrix flows.*

(iv_{2b}) For $m = 2m_1$ and $n = 2n_1 + 1$, the equilibriums of $(x^*_{j_1}, x^*_{j_2}) = (a_{j_1j_1\alpha}, a_{j_1j_2\beta})$ $(\alpha \in \{1, 2, \ldots, 2m_1\}, \beta \in \{1, 2, \ldots, 2n_1 + 1\})$ for the x_{j_1}-direction are a hyperbolic and hyperbolic-secant flow network as follows as in Eqs. (1.70–1.73).

- *For $a_{j_1j_10} > 0$ and $a_{j_2j_20} > 0$,*

$$\bigcup_{\alpha=1}^{2m_1}\bigcup_{\beta=1}^{2n_1+1}\underbrace{(a_{j_1j_1\alpha}, a_{j_1j_2\beta})}_{\text{(XX, pF)}}$$

$$= \begin{bmatrix} \underbrace{\text{(UP:DP,pF)}}_{\text{hyperbolic-secant flow (+)}} & \cdots & \underbrace{\text{(UP:DP,pF)}}_{\text{hyperbolic-secant flow (+)}} & \underbrace{\text{(DP:UP,pF)}}_{\text{hyperbolic flow (+)}} \\ \underbrace{\text{(DP:UP,pF)}}_{\text{hyperbolic flow (+)}} & \cdots & \underbrace{\text{(DP:UP,pF)}}_{\text{hyperbolic flow (+)}} & \underbrace{\text{(UP:DP,pF)}}_{\text{hyperbolic-secant flow (+)}} \\ \vdots & \vdots & \vdots & \vdots \\ \underbrace{\text{(UP:DP,pF)}}_{\text{hyperbolic-secant flow (+)}} & \cdots & \underbrace{\text{(UP:DP,pF)}}_{\text{hyperbolic-secant flow (+)}} & \underbrace{\text{(DP:UP,pF)}}_{\text{hyperbolic flow (+)}} \end{bmatrix}_{(2n_1+1)\times(2m_1)}.$$

- *For $a_{j_1j_10} < 0$ and $a_{j_2j_20} > 0$,*

$$\bigcup_{\alpha=1}^{2m_1}\bigcup_{\beta=1}^{2n_1+1}\underbrace{(a_{j_1j_1\alpha}, a_{j_1j_2\beta})}_{\text{(XX, pF)}}$$

$$= \begin{bmatrix} \underbrace{\text{(DP:UP,pF)}}_{\text{hyperbolic flow (+)}} & \cdots & \underbrace{\text{(DP:UP,pF)}}_{\text{hyperbolic flow (+)}} & \underbrace{\text{(UP:DP,pF)}}_{\text{hyperbolic-secant flow (+)}} \\ \underbrace{\text{(UP:DP,pF)}}_{\text{hyperbolic-secant flow (+)}} & \cdots & \underbrace{\text{(UP:DP,pF)}}_{\text{hyperbolic-secant flow (+)}} & \underbrace{\text{(DP:UP,pF)}}_{\text{hyperbolic flow (+)}} \\ \vdots & \vdots & \vdots & \vdots \\ \underbrace{\text{(DP:UP,pF)}}_{\text{hyperbolic flow (+)}} & \cdots & \underbrace{\text{(DP:UP,pF)}}_{\text{hyperbolic flow (+)}} & \underbrace{\text{(UP:DP,pF)}}_{\text{hyperbolic-secant flow (+)}} \end{bmatrix}_{(2n_1+1)\times(2m_1)}.$$

- *For $a_{j_1j_10} > 0$ and $a_{j_2j_20} < 0$,*

$$\bigcup_{\alpha=1}^{2m_1}\bigcup_{\beta=1}^{2n_1+1}\underbrace{(a_{j_1j_1\alpha}, a_{j_1j_2\beta})}_{\text{(XX, nF)}} = \begin{bmatrix} \underbrace{\text{(DP:UP,nF)}}_{\text{hyperbolic flow }(-)} & \cdots & \underbrace{\text{(DP:UP,nF)}}_{\text{hyperbolic flow }(-)} & \underbrace{\text{(UP:DP,nF)}}_{\text{hyperbolic-secant flow }(-)} \\ \underbrace{\text{(UP:DP,nF)}}_{\text{hyperbolic-secant flow }(-)} & \cdots & \underbrace{\text{(UP:DP,nF)}}_{\text{hyperbolic-secant flow }(-)} & \underbrace{\text{(DP:UP,nF)}}_{\text{hyperbolic flow }(-)} \\ \vdots & \vdots & \vdots & \vdots \\ \underbrace{\text{(DP:UP,nF)}}_{\text{hyperbolic flow }(-)} & \cdots & \underbrace{\text{(DP:UP,nF)}}_{\text{hyperbolic flow }(-)} & \underbrace{\text{(UP:DP,nF)}}_{\text{hyperbolic-secant flow }(-)} \end{bmatrix}_{(2n_1+1)\times(2m_1)}.$$

- *For $a_{j_1j_10} < 0$ and $a_{j_2j_20} < 0$,*

$$\bigcup_{\alpha=1}^{2m_1}\bigcup_{\beta=1}^{2n_1+1}\underbrace{(a_{j_1j_1\alpha}, a_{j_1j_2\beta})}_{\text{(XX, nF)}} = \begin{bmatrix} \underbrace{\text{(UP:DP,nF)}}_{\text{hyperbolic-secant flow }(-)} & \cdots & \underbrace{\text{(UP:DP,nF)}}_{\text{hyperbolic-secant flow }(-)} & \underbrace{\text{(DP:UP,nF)}}_{\text{hyperbolic flow }(-)} \\ \underbrace{\text{(DP:UP,nF)}}_{\text{hyperbolic flow }(-)} & \cdots & \underbrace{\text{(DP:UP,nF)}}_{\text{hyperbolic flow }(-)} & \underbrace{\text{(UP:DP,nF)}}_{\text{hyperbolic-secant flow }(-)} \\ \vdots & \vdots & \vdots & \vdots \\ \underbrace{\text{(UP:DP,nF)}}_{\text{hyperbolic-secant flow }(-)} & \cdots & \underbrace{\text{(UP:DP,nF)}}_{\text{hyperbolic-secant flow }(-)} & \underbrace{\text{(DP:UP,nF)}}_{\text{hyperbolic flow }(-)} \end{bmatrix}_{(2n_1+1)\times(2m_1)}.$$

The lines of $x^*_{j_1} = a_{j_1j_1\alpha}$ ($\alpha \in \{1, 2, \ldots, 2m_1\}$) are directrix flows.

(iv_{2c}) For $m = 2m_1 + 1$ and $n = 2n_1$, the equilibriums of $(x^*_{j_1}, x^*_{j_2}) = (a_{j_1j_1\alpha}, a_{j_1j_2\beta})$ ($\alpha \in \{1, 2, \ldots, 2m_1 + 1\}, \beta \in \{1, 2, \ldots, 2n_1\}$) for the x_{j_1}-direction are a hyperbolic and hyperbolic-secant flow network as in Eqs. (1.74–1.77).

- *For* $a_{j_1j_10} > 0$ *and* $a_{j_2j_20} > 0$,

$$
\bigcup_{\alpha=1}^{2m_1+1}\bigcup_{\beta=1}^{2n_1}\underbrace{(a_{j_1j_1\alpha}, a_{j_1j_2\beta})}_{(\mathrm{XX},\ \mathrm{pF})}
= \begin{bmatrix}
\underbrace{(\mathrm{DP{:}UP,pF})}_{\text{hyperbolic flow }(+)} & \cdots & \underbrace{(\mathrm{UP{:}DP,pF})}_{\text{hyperbolic-secant flow }(+)} & \underbrace{(\mathrm{DP{:}UP,pF})}_{\text{hyperbolic flow }(+)} \\
\underbrace{(\mathrm{UP{:}DP,pF})}_{\text{hyperbolic-secant flow }(+)} & \cdots & \underbrace{(\mathrm{DP{:}UP,pF})}_{\text{hyperbolic flow }(+)} & \underbrace{(\mathrm{UP{:}DP,pF})}_{\text{hyperbolic-secant flow }(+)} \\
\vdots & \vdots & \vdots & \vdots \\
\underbrace{(\mathrm{UP{:}DP,pF})}_{\text{hyperbolic-secant flow }(+)} & \cdots & \underbrace{(\mathrm{DP{:}UP,pF})}_{\text{hyperbolic flow }(+)} & \underbrace{(\mathrm{UP{:}DP,pF})}_{\text{hyperbolic-secant flow }(+)}
\end{bmatrix}_{(2n_1)\times(2m_1+1)}.
$$

- *For* $a_{j_1j_10} < 0$ *and* $a_{j_2j_20} > 0$,

$$
\bigcup_{\alpha=1}^{2m_1+1}\bigcup_{\beta=1}^{2n_1}\underbrace{(a_{j_1j_1\alpha}, a_{j_1j_2\beta})}_{(\mathrm{XX},\ \mathrm{pF})}
= \begin{bmatrix}
\underbrace{(\mathrm{UP{:}DP,pF})}_{\text{hyperbolic-secant flow }(+)} & \cdots & \underbrace{(\mathrm{DP{:}UP,pF})}_{\text{hyperbolic flow }(+)} & \underbrace{(\mathrm{UP{:}DP,pF})}_{\text{hyperbolic-secant flow }(+)} \\
\underbrace{(\mathrm{DP{:}UP,pF})}_{\text{hyperbolic flow }(+)} & \cdots & \underbrace{(\mathrm{UP{:}DP,pF})}_{\text{hyperbolic-secant flow }(+)} & \underbrace{(\mathrm{DP{:}UP,pF})}_{\text{hyperbolic flow }(+)} \\
\vdots & \vdots & \vdots & \vdots \\
\underbrace{(\mathrm{DP{:}UP,pF})}_{\text{hyperbolic flow }(+)} & \cdots & \underbrace{(\mathrm{UP{:}DP,pF})}_{\text{hyperbolic-secant flow }(+)} & \underbrace{(\mathrm{DP{:}UP,pF})}_{\text{hyperbolic flow }(+)}
\end{bmatrix}_{(2n_1)\times(2m_1+1)}.
$$

- *For* $a_{j_1j_10} > 0$ *and* $a_{j_2j_20} < 0$,

$$
\bigcup_{\alpha=1}^{2m_1+1}\bigcup_{\beta=1}^{2n_1}\underbrace{(a_{j_1j_1\alpha}, a_{j_1j_2\beta})}_{(\mathrm{XX},\ \mathrm{nF})}
= \begin{bmatrix}
\underbrace{(\mathrm{UP{:}DP,nF})}_{\text{hyperbolic-secant flow }(-)} & \cdots & \underbrace{(\mathrm{DP{:}UP,nF})}_{\text{hyperbolic flow }(-)} & \underbrace{(\mathrm{UP{:}DP,nF})}_{\text{hyperbolic-secant flow }(-)} \\
\underbrace{(\mathrm{DP{:}UP,nF})}_{\text{hyperbolic flow }(-)} & \cdots & \underbrace{(\mathrm{UP{:}DP,nF})}_{\text{hyperbolic-secant flow }(-)} & \underbrace{(\mathrm{DP{:}UP,nF})}_{\text{hyperbolic flow }(-)} \\
\vdots & \vdots & \vdots & \vdots \\
\underbrace{(\mathrm{DP{:}UP,nF})}_{\text{hyperbolic flow }(-)} & \cdots & \underbrace{(\mathrm{UP{:}DP,nF})}_{\text{hyperbolic-secant flow }(-)} & \underbrace{(\mathrm{DP{:}UP,nF})}_{\text{hyperbolic flow }(-)}
\end{bmatrix}_{(2n_1)\times(2m_1+1)}.
$$

- *For $a_{j_1j_10} < 0$ and $a_{j_2j_20} < 0$,*

$$\bigcup_{\alpha=1}^{2m_1+1}\bigcup_{\beta=1}^{2n_1}\underbrace{(a_{j_1j_1\alpha},a_{j_1j_2\beta})}_{\text{(XX, nF)}}$$

$$=\begin{bmatrix} \underbrace{\text{(DP:UP,nF)}}_{\text{hyperbolic flow }(-)} & \cdots & \underbrace{\text{(UP:DP,nF)}}_{\text{hyperbolic-secant flow }(-)} & \underbrace{\text{(DP:UP,nF)}}_{\text{hyperbolic flow }(-)} \\ \underbrace{\text{(UP:DP,nF)}}_{\text{hyperbolic-secant flow }(-)} & \cdots & \underbrace{\text{(DP:UP,nF)}}_{\text{hyperbolic flow }(-)} & \underbrace{\text{(UP:DP,nF)}}_{\text{hyperbolic-secant flow }(-)} \\ \vdots & \vdots & \vdots & \vdots \\ \underbrace{\text{(UP:DP,nF)}}_{\text{hyperbolic-secant flow }(-)} & \cdots & \underbrace{\text{(DP:UP,nF)}}_{\text{hyperbolic flow }(-)} & \underbrace{\text{(UP:DP,nF)}}_{\text{hyperbolic-secant flow }(-)} \end{bmatrix}_{(2n_1)\times(2m_1+1)}.$$

The lines of $x_{j_1}^* = a_{j_1j_1\alpha}$ $(\alpha \in \{1, 2, \ldots, 2m_1 + 1\})$ are directrix flows.

(iv_{2d}) For $m = 2m_1$ and $n = 2n_1$, the equilibriums of $(x_{j_1}^*, x_{j_2}^*) = (a_{j_1j_1\alpha}, a_{j_1j_2\beta})$ $(\alpha \in \{1, 2, \ldots, 2m_1\}, \beta \in \{1, 2, \ldots, 2n_1\})$ for the x_{j_1}-direction are a hyperbolic and hyperbolic-secant flow network as follows as in Eqs. (1.78–1.81).

- *For $a_{j_1j_10} > 0$ and $a_{j_2j_20} > 0$,*

$$\bigcup_{\alpha=1}^{2m_1}\bigcup_{\beta=1}^{2n_1}\underbrace{(a_{j_1j_1\alpha},a_{j_1j_2\beta})}_{\text{(XX, pF)}}$$

$$=\begin{bmatrix} \underbrace{\text{(UP:DP,pF)}}_{\text{hyperbolic-secant flow }(+)} & \cdots & \underbrace{\text{(UP:DP,pF)}}_{\text{hyperbolic-secant flow }(+)} & \underbrace{\text{(DP:UP,pF)}}_{\text{hyperbolic flow }(+)} \\ \underbrace{\text{(DP:UP,pF)}}_{\text{hyperbolic flow }(+)} & \cdots & \underbrace{\text{(DP:UP,pF)}}_{\text{hyperbolic flow }(+)} & \underbrace{\text{(UP:DP,pF)}}_{\text{hyperbolic-secant flow }(+)} \\ \vdots & \vdots & \vdots & \vdots \\ \underbrace{\text{(DP:UP,pF)}}_{\text{hyperbolic flow }(+)} & \cdots & \underbrace{\text{(DP:UP,pF)}}_{\text{hyperbolic flow }(+)} & \underbrace{\text{(UP:DP,pF)}}_{\text{hyperbolic-secant flow }(+)} \end{bmatrix}_{(2n_1)\times(2m_1)}.$$

- *For* $a_{j_1j_10} < 0$ *and* $a_{j_2j_20} > 0$,

$$\bigcup_{\alpha=1}^{2m_1}\bigcup_{\beta=1}^{2n_1}\underbrace{(a_{j_1j_1\alpha}, a_{j_1j_2\beta})}_{(\text{XX, pF})}$$
$$=\begin{bmatrix}
\underbrace{\text{(DP:UP,pF)}}_{\text{hyperbolic flow }(+)} & \cdots & \underbrace{\text{(DP:UP,pF)}}_{\text{hyperbolic flow }(+)} & \underbrace{\text{(UP:DP,pF)}}_{\text{hyperbolic-secant flow }(+)} \\
\underbrace{\text{(UP:DP,pF)}}_{\text{hyperbolic-secant flow }(+)} & \cdots & \underbrace{\text{(UP:DP,pF)}}_{\text{hyperbolic-secant flow }(+)} & \underbrace{\text{(DP:UP,pF)}}_{\text{hyperbolic flow }(+)} \\
\vdots & \vdots & \vdots & \vdots \\
\underbrace{\text{(UP:DP,pF)}}_{\text{hyperbolic-secant flow }(+)} & \cdots & \underbrace{\text{(UP:DP,pF)}}_{\text{hyperbolic-secant flow }(+)} & \underbrace{\text{(DP:UP,pF)}}_{\text{hyperbolic flow }(+)}
\end{bmatrix}_{(2n_1)\times(2m_1)}.$$

- *For* $a_{j_1j_10} > 0$ *and* $a_{j_2j_20} < 0$,

$$\bigcup_{\alpha=1}^{2m_1}\bigcup_{\beta=1}^{2n_1}\underbrace{(a_{j_1j_1\alpha}, a_{j_1j_2\beta})}_{(\text{XX, nF})}$$
$$=\begin{bmatrix}
\underbrace{\text{(DP:UP,nF)}}_{\text{hyperbolic flow }(-)} & \cdots & \underbrace{\text{(DP:UP,nF)}}_{\text{hyperbolic flow }(-)} & \underbrace{\text{(UP:DP,nF)}}_{\text{hyperbolic-secant flow }(-)} \\
\underbrace{\text{(UP:DP,nF)}}_{\text{hyperbolic-secant flow }(-)} & \cdots & \underbrace{\text{(UP:DP,nF)}}_{\text{hyperbolic-secant flow }(-)} & \underbrace{\text{(DP:UP,nF)}}_{\text{hyperbolic flow }(-)} \\
\vdots & \vdots & \vdots & \vdots \\
\underbrace{\text{(UP:DP,nF)}}_{\text{hyperbolic-secant flow }(-)} & \cdots & \underbrace{\text{(UP:DP,nF)}}_{\text{hyperbolic-secant flow }(-)} & \underbrace{\text{(DP:UP,nF)}}_{\text{hyperbolic flow }(-)}
\end{bmatrix}_{(2n_1)\times(2m_1)}.$$

- *For* $a_{j_1j_10} < 0$ *and* $a_{j_2j_20} < 0$,

$$\bigcup_{\alpha=1}^{2m_1}\bigcup_{\beta=1}^{2n_1}\underbrace{(a_{j_1j_1\alpha}, a_{j_1j_2\beta})}_{(\text{XX, nF})}$$
$$=\begin{bmatrix}
\underbrace{\text{(UP:DP,nF)}}_{\text{hyperbolic-secant flow }(-)} & \cdots & \underbrace{\text{(UP:DP,nF)}}_{\text{hyperbolic-secant flow }(-)} & \underbrace{\text{(DP:UP,nF)}}_{\text{hyperbolic flow }(-)} \\
\underbrace{\text{(DP:UP,nF)}}_{\text{hyperbolic flow }(-)} & \cdots & \underbrace{\text{(DP:UP,nF)}}_{\text{hyperbolic flow }(-)} & \underbrace{\text{(UP:DP,nF)}}_{\text{hyperbolic-secant flow }(-)} \\
\vdots & \vdots & \vdots & \vdots \\
\underbrace{\text{(DP:UP,nF)}}_{\text{hyperbolic flow }(-)} & \cdots & \underbrace{\text{(DP:UP,nF)}}_{\text{hyperbolic flow }(-)} & \underbrace{\text{(UP:DP,nF)}}_{\text{hyperbolic-secant flow }(-)}
\end{bmatrix}_{(2n_1)\times(2m_1)}.$$

The lines of $x_{j_1}^* = a_{j_1j_1\alpha}$ $(\alpha \in \{1, 2, \ldots, 2m_1\})$ *are directrix flows.*

2.5 Singular Flows and Bifurcations

(v) *For* $m = 2m_1 + 1$ *and* $n = 2n_1 + 1$, *the equilibrium of* $(x^*_{j_1}, x^*_{j_2}) = (a_{j_1j_1 1}, a_{j_1j_2 1})$ *for the* x_{j_1}*-direction has the following bifurcation properties with*

$$
\begin{aligned}
&2m_1 + 1 - \sum_{\alpha_1=1}^{p_1} m_{\alpha_1} = 2\sum_{\rho_1=1}^{l_1} m_{\rho_1},\ 2n_1 + 1 - \sum_{\beta_1=1}^{q_1} n_{\beta_1} = 2\sum_{\sigma_1=1}^{l_2} n_{\sigma_1};\\
&\sum_{\alpha_2=1}^{p_2} m_{\alpha_2} = 2m_1 + 1,\ \sum_{\beta_2=1}^{q_2} n_{\beta_2} = 2n_1 + 1;\\
&\sum_{\alpha_i=1}^{p} m_{\alpha_i} = 2m_1 + 1,\ \sum_{\beta_i=1}^{q} n_{\beta_i} = 2n_1 + 1,\\
&\text{for } i = 1, 2;
\end{aligned}
$$

l_1, l_2-quadratic polynomials are without real roots and ρ_1^{th} and σ_1^{th}-quadratic polynomials are with powers m_{ρ_1} and n_{σ_1},respectively.

The bifurcation process is discussed through differential equations as follows. There are four bifurcations (I_1)–(I_3).

(I_1) Consider a dynamical system having a single equilibrium (A_1) as

$$
\begin{aligned}
\dot{x}_{j_1} &= a_{j_1j_1 0}(x_{j_1} - b_{j_1j_1 1}) \prod_{r=1}^{m_1} [(x_{j_1} - a^{(r)}_{j_1j_1})^2 + \Delta^{(r)}_{j_1j_1}]\\
&\quad \times (x_{j_2} - b_{j_1j_2 1}) \prod_{s=1}^{n_1} [(x_{j_2} - a^{(s)}_{j_1j_2})^2 + \Delta^{(s)}_{j_1j_2}],\\
\dot{x}_{j_2} &= a_{j_2j_1 0};
\end{aligned}
$$

where $a_{j_1j_1 1} = b_{j_1j_1 1}$ and $a_{j_2j_1 1} = b_{j_2j_1 1}$. Once

$$
\begin{aligned}
\Delta^{(r)}_{j_1j_1} &= \Delta_{j_1j_1} = 0\,(r = 1, 2, \ldots, m_1),\\
\Delta^{(s)}_{j_1j_2} &= \Delta_{j_1j_2} = 0\,(s = 1, 2, \ldots, n_1),
\end{aligned}
$$

if

$$
\begin{aligned}
a_{j_1j_1 1} &= b_{j_1j_1 1} = a^{(r)}_{j_1j_1}\ (r = 1, 2, \ldots, m_1),\\
a_{j_1j_2 1} &= b_{j_1j_2 1} = a^{(s)}_{j_1j_2}\ (s = 1, 2, \ldots, n_1)
\end{aligned}
$$

the foregoing differential equations becomes a dynamical system with a singular equilibrium (B_1) as

$$\dot{x}_{j_1} = a_{j_1j_10}(x_{j_1} - a_{j_1j_11})^{2m_1+1}(x_{j_2} - a_{j_1j_21})^{2n_1+1}, \ \dot{x}_{j_2} = a_{j_2j_10}.$$

If

$$\Delta_{j_1j_1}^{(r)} = \Delta_{j_1j_1} + \delta_r \ (r = 1, 2, \ldots, m_1), \ \delta_r > 0;$$
$$\Delta_{j_1j_2}^{(s)} = \Delta_{j_1j_2} + \delta_s \ (s = 1, 2, \ldots, n_1), \ \delta_s > 0;$$

then

$$a_{j_1j_11}^{(r)}, a_{j_1j_12}^{(r)} = a_{j_1j_11} \pm \varepsilon_r, \ (r = 1, 2, \ldots, m_1),$$
$$\{a_{j_1j_11}, a_{j_1j_12}, \ldots, a_{j_1j_1(2m_1+1)}\} = \text{sort}\{\ b_{j_1j_11}, a_{j_1j_11}^{(r)}, a_{j_1j_12}^{(r)} | r = 1, 2, \ldots, m_1\}\ ;$$
$$a_{j_1j_21}^{(s)}, a_{j_1j_22}^{(s)} = a_{j_1j_21} \pm \varepsilon_s, \ (s = 1, 2, \ldots, n_1),$$
$$\{a_{j_1j_21}, a_{j_1j_22}, \ldots, a_{j_1j_2(2n_1+1)}\} = \text{sort}\{\ b_{j_1j_21}, a_{j_1j_21}^{(s)}, a_{j_1j_22}^{(s)} | s = 1, 2, \ldots, n_1\}\ .$$

Thus, the differential equation becomes a dynamical system with a non-singular equilibrium network (C_1) as

$$\dot{x}_{j_1} = a_{j_1j_10} \prod_{\alpha=1}^{2m_1+1} (x_{j_1} - a_{j_1j_1\alpha}) \prod_{\beta=1}^{2n_1+1} (x_{j_2} - a_{j_1j_2\beta}), \ \dot{x}_{j_2} = a_{j_2j_10}.$$

- From the above bifurcation process analysis, at least, only (m_1+1)-parameter variations in the x_{j_1}-direction are engaged in such a bifurcation from a 1-dimensional flow to the 1-dimensional flow array of $1 \times (2m_1 + 1)$ through the $(2m_1 + 1)^{\text{th}}$-source and sink bifurcations in the x_{j_1}-direction. *The higher-order source and sink flow bifurcations are developed.*

  Thus, the 1-dimensional flow appearing or vanishing bifurcation route is as follows.

$$\underbrace{(a_{j_1j_11}, \dot{x}_{j_2})}_{\text{XX}} \rightleftharpoons \underbrace{(a_{j_1j_11}, \dot{x}_{j_2})}_{(2m_1+1)^{\text{th}}\text{order-XX}} \rightleftharpoons \bigcup_{r=1}^{2m_1+1} \underbrace{(a_{j_1j_1r}, \dot{x}_{j_2})}_{\text{XX}}.$$

- From the above bifurcation process analysis, at least, only (n_1+1)-parameter variations in the x_{j_1}-direction are engaged in such a bifurcation from a 1-dimensional flow to the 1-dimensional flow array of $(2n_1 + 1) \times 1$ through the $(2n_1 + 1)^{\text{th}}$-up-parabola and down-parabola bifurcations in the x_{j_1}-direction. *The higher-order up-parabola and lower-parabola flow bifurcations are developed.*

  Thus, the 1-dimensional flow appearing or vanishing bifurcation route is as follows.

$$\underbrace{(a_{j_1j_21},\dot{x}_{j_2})}_{\text{YY}} \rightleftharpoons \underbrace{(a_{j_1j_21},\dot{x}_{j_2})}_{(2n_1+1)^{\text{th}}\text{order-YY}} \rightleftharpoons \bigcup_{\beta=1}^{2n_1+1} \underbrace{(a_{j_1j_2\beta},\dot{x}_{j_2})}_{\text{YY}}.$$

- From the above bifurcation process analysis, (m_1+n_1+2)-parameter variations in the two directions are engaged in such a bifurcation from a parabola-flow to a $(2m_1+1)\times(2n_1+1)$ network of hyperbolic and hyperbolic-secant flows through the $(2m_1+1)^{th}$-source and sink bifurcations in the x_{j_1}-direction, and the $(2n_1+1)^{\text{th}}$-up-parabola and down-parabola flows bifurcations in the x_{j_1}-direction. *The higher-order hyperbolic and hyperbolic-recant flow bifurcations are developed.*

  Thus, the 1-dimensional flow appearing or vanishing bifurcation route is as follows.

$$\begin{aligned}&\underbrace{(a_{j_1j_11},a_{j_1j_21})}_{\text{(XX,xF)-flow}} \rightleftharpoons \underbrace{[(2m_1+1),(2n_1+1)]\text{XX},xF)}_{[(2m_1+1),(2n_1+1)]\text{-XX}}\\ &\rightleftharpoons \bigcup_{\alpha=1}^{2m_1+1}\bigcup_{\beta=1}^{2n_1+1} \underbrace{(a_{j_1j_1\alpha},a_{j_1j_2\beta})}_{\text{(YY,xF)-flow}}.\end{aligned}$$

(I_2) From Case (I_1), consider a dynamical system having singular equilibriums (A_2) as

$$\begin{aligned}\dot{x}_{j_1} &= a_{j_1j_10}\prod_{r_1}^{p_1}(x_{j_1}-b_{j_1j_1r_1})^{m_{r_1}}\prod_{r=1}^{l_1}[(x_{j_1}-a_{j_1j_1}^{(r)})^2+\Delta_{j_1j_1}^{(r)}]^{m_r}\\ &\quad\times\prod_{r_1}^{q_1}(x_{j_2}-b_{j_1j_2r_1})^{m_{r_1}}\prod_{s=1}^{l_2}[(x_{j_2}-a_{j_1j_2}^{(s)})^2+\Delta_{j_1j_2}^{(s)}]^{m_s},\\ \dot{x}_{j_2} &= a_{j_2j_10};\end{aligned}$$

where

$$\begin{aligned}&2m_1+1-\sum_{\alpha_1=1}^{p_1}m_{\alpha_1}=2\sum_{\rho_1=1}^{l_1}m_{\rho_1},\ 2n_1+1-\sum_{\beta_1=1}^{q_1}n_{\beta_1}=2\sum_{\sigma_1=1}^{l_2}n_{\sigma_1};\\ &\sum_{\alpha_2=1}^{p_2}m_{\alpha_2}=2m_1+1,\ \sum_{\beta_2=1}^{q_2}n_{\beta_2}=2n_1+1;\end{aligned}$$

l_1,l_2-quadratic polynomials are without real roots and ρ_1^{th} and σ_1^{th}-quadratic polynomials are with powers m_{ρ_1} and n_{σ_1},respectively.

Once

$$\Delta_{j_1 j_1}^{(r)} = 0 \, (r = 1, 2, \ldots, l_1),$$
$$\Delta_{j_1 j_2}^{(s)} = 0 \, (s = 1, 2, \ldots, l_2);$$

if

$$a_{j_1 j_1 1} = a_{j_1 j_1 r_1} = a_{j_1 j_1}^{(r)} \, (r_1 = 1, 2, \ldots, p_1; r = 1, 2, \ldots, l_1),$$
$$a_{j_1 j_2 1} \equiv a_{j_1 j_2 r_1} = a_{j_1 j_2}^{(s)} \, (s_1 = 1, 2, \ldots, q_1; s = 1, 2, \ldots, l_2),$$

the foregoing differential equations becomes a dynamical system with a 1-dimensional flow (B_2) as

$$\dot{x}_{j_1} = a_{j_1 j_1 0}(x_{j_1} - a_{j_1 j_1 1})^{2m_1+1}(x_{j_2} - a_{j_1 j_2 1})^{2n_1+1}, \; \dot{x}_{j_2} = a_{j_2 j_1 0}.$$

If

$$\Delta_{j_1 j_1}^{(r)} = \Delta_{j_1 j_1} + \delta_r \, (r = 1, 2, \ldots, l_1), \; \delta_r > 0;$$
$$\Delta_{j_1 j_2}^{(s)} = \Delta_{j_1 j_2} + \delta_s \, (s = 1, 2, \ldots, l_2), \; \delta_s > 0;$$

then

$$\begin{aligned}
&a_{j_1 j_1 1}^{(r)}, a_{j_1 j_1 2}^{(r)} = a_{j_1 j_1 1} \pm \varepsilon_r, \; (r = 1, 2, \ldots, l_1), \\
&\left\{a_{j_1 j_1 1}, a_{j_1 j_1 2}, \ldots, a_{j_1 j_1 p_2}\right\} \\
&= \text{sort}\left\{a_{j_1 j_1 1}, a_{j_1 j_1 2}, \ldots, a_{j_1 j_1 p_1}; a_{j_1 j_1 1}^{(r)}, a_{j_1 j_1 2}^{(r)} | r = 1, 2, \ldots, l_1 \,;\right\} \\
&a_{j_1 j_2 1}^{(s)}, a_{j_1 j_2 2}^{(s)} = a_{j_1 j_2 1} \pm \varepsilon_s, \; (s = 1, 2, \ldots, l_2), \\
&\left\{a_{j_1 j_2 1}, a_{j_1 j_2 2}, \ldots, a_{j_1 j_2 q_2}\right\} \\
&= \text{sort}\left\{a_{j_1 j_2 1}, a_{j_1 j_2 2}, \ldots, a_{j_1 j_2 q_1}; a_{j_1 j_2 1}^{(s)}, a_{j_1 j_2 2}^{(s)} | s = 1, 2, \ldots, l_2 \right\}.
\end{aligned}$$

Thus, the differential equation becomes a dynamical system with singular 1-dimensional flows (C_2) as

$$\begin{aligned}
\dot{x}_{j_1} &= a_{j_1 j_1 0} \prod_{\alpha_2}^{p_2} (x_{j_1} - a_{j_1 j_1 \alpha_2})^{m_{\alpha 2}} \prod_{\beta_2}^{q_2} (x_{j_2} - a_{j_1 j_2 \beta_2})^{n_{\beta_2}}, \\
\dot{x}_{j_2} &= a_{j_2 j_1 0}
\end{aligned}$$

where

$$\sum_{\alpha_2=1}^{p_2} m_{\alpha_2}^{(2)} = 2m_1 + 1, \sum_{\beta_2=1}^{q_2} n_{\beta_2}^{(2)} = 2n_1 + 1.$$

- From the above bifurcation process analysis, only p_2-parameter variations in the x_{j_1}-direction from a p_1 to p_2 1-dimensional flow array through the $(2m_1+1)^{\text{th}}$-source and sink flow bifurcations in the x_{j_1}-direction.

  Thus, the 1-dimensional flow appearing or vanishing bifurcation route from a p_1 to p_2 1-dimensional flow array is expressed as

$$\bigcup_{r_1=1}^{p_1} \underbrace{(a_{j_1j_1r_1}, \dot{x}_{j_2})}_{(\mathrm{m}_{r_1}^{(1)})^{\text{th}}\text{order XX}} \rightleftharpoons \underbrace{(a_{j_1j_1 1}, \dot{x}_{j_2})}_{(2m_1+1)^{\text{th}}\text{ order XX}} \rightleftharpoons \bigcup_{r_2=1}^{p_2} \underbrace{(a_{j_1j_1r_2}, \dot{x}_{j_2})}_{(\mathrm{m}_{r_2}^{(2)})^{\text{th}}\text{order XX}}.$$

- From the above bifurcation process analysis, only q_2-parameter variations in the x_{j_1}-direction from a q_1 to q_2 1-dimensional flow array through the $(2n_1+1)^{\text{th}}$-up-parabola and down-parabola flow bifurcations in the x_{j_1}-direction.

  Thus, the 1-dimensional flow appearing or vanishing bifurcation route from a q_1 to q_2 1-dimensional flow array is expressed as

$$\bigcup_{\beta_1=1}^{q_1} \underbrace{(a_{j_2j_1\beta_1}, \dot{x}_{j_2})}_{(n_{\beta_1}^{(1)})^{\text{th}}\text{order YY}} \rightleftharpoons \underbrace{(a_{j_1j_2 1}, \dot{x}_{j_2})}_{(2n_1+1)^{\text{th}}\text{ order YY}} \rightleftharpoons \bigcup_{s_2=1}^{q_2} \underbrace{(a_{j_1j_2\beta_2}, \dot{x}_{j_2})}_{(n_{\beta_2}^{(2)})^{\text{th}}\text{order YY}}.$$

- From the above bifurcation process analysis, only q_2+p_2-parameter variations in the x_{j_1}-direction from a $q_1\times p_1$ to $q_2\times p_2$ network of 1-dimensional flows through the $(2m_1+1)^{\text{th}}$-source and sink flows bifurcations in the x_{j_1}-direction, and the $(2n_1+1)^{\text{th}}$-up-parabola and down-parabola flows bifurcations in the x_{j_1}-direction.

  Thus, the 1-dimensional flow appearing or vanishing bifurcation route is as follows:

$$\begin{aligned}
&\bigcup_{\alpha_1=1}^{p_1}\bigcup_{\beta_1=1}^{q_1} \underbrace{(a_{j_1j_1\alpha_1}, a_{j_1j_2\beta_1})}_{([(m_{r_1}),(n_{s_1})]\text{XX,xF)-flow}} \rightleftharpoons \underbrace{[(2m_1+1),\ (2n_1+1)]\text{XX, xF)}}_{[(2m_1+1),(2n_1+1)]\text{-XX}}\\
&\rightleftharpoons \bigcup_{\alpha_2=1}^{p_2}\bigcup_{\beta_2=1}^{q_2} \underbrace{(a_{j_1j_1\alpha_2}, a_{j_1j_2\beta_2})}_{([(m_{r_1}),(n_{s_1})]\text{XX,xF)-flow}}.
\end{aligned}$$

(I_3) Consider two dynamical systems with the same 1-dimensional flows with locations switched (A_3,C_3) for $i=1,2$ as

$$\begin{aligned}
\dot{x}_{j_1} &= a_{j_1j_1 0}\prod_{r_i}^{p}(x_{j_1}-a_{j_1j_1r_i})^{m_{r_i}^{(i)}}\prod_{s_i}^{q}(x_{j_1}-a_{j_2j_1s_i})^{n_{s_i}^{(i)}},\\
\dot{x}_{j_2} &= a_{j_2j_1 0};
\end{aligned}$$

where

$$\sum_{r_i=1}^{p_2} m_{r_i}^{(i)} = 2m_1 + 1, \sum_{\beta_2=1}^{q_2} n_{\beta_2}^{(2)} = 2n_1 + 1.$$

Consider a dynamical system as a singular 1-dimensional flow (B_3) as

$$\dot{x}_{j_1} = a_{j_1 j_1 0}(x_{j_1} - a_{j_1 j_1 1})^{2m_1+1}(x_{j_2} - a_{j_1 j_2 1})^{2n_1+1}, \ \dot{x}_{j_2} = a_{j_2 j_1 0}.$$

For $r_i = 1, 2, \ldots, p$ and $s_i = 1, 2, \ldots, q$ $(i = 1, 2)$, define a function as

$$\begin{aligned}\Delta_{j_1 j_1}^{(r_1 r_2)} &= (a_{j_1 j_1 r_1} - a_{j_1 j_1 r_2})^2 \ (r_1, r_2 \in \{1, 2, \ldots, p\}, r_1 \neq r_2),\\ \Delta_{j_1 j_2}^{(s_1 s_2)} &= (a_{j_1 j_2 s_1} - a_{j_1 j_2 s_2})^2 \ (s_1, s_2 \in \{1, 2, \ldots, q\}, s_1 \neq s_2).\end{aligned}$$

For $r_i = 1, 2, \ldots, p$ and $s_i = 1, 2, \ldots, q$ $(i = 1, 2)$, if

$$\Delta_{j_1 j_1}^{(r_1 r_2)} = (a_{j_1 j_1 r_1} - a_{j_1 j_1 r_2})^2 = 0 \text{ and } \Delta_{j_1 j_2}^{(s_1 s_2)} = (a_{j_1 j_2 s_1} - a_{j_1 j_2 s_2})^2 = 0,$$

- Two 1-dimensional flows of $(a_{j_1 j_1 r_1}, \dot{x}_{j_2})$ and $(a_{j_1 j_1 r_2}, \dot{x}_{j_2})$ switching at point $(a_{j_1 j_1 1}, \dot{x}_{j_2})$ with the same order singularity are given through

$$\bigcup_{r_1=1}^{p} \underbrace{(a_{j_1 j_1 r_1}, \dot{x}_{j_2})}_{(m_{r_1}^{(1)})^{\text{th}}\text{order XX}} \rightleftharpoons \underbrace{(a_{j_1 j_1 1}, \dot{x}_{j_2})}_{(2m_1+1)^{\text{th}} \text{ order XX}} \rightleftharpoons \bigcup_{r_2=1}^{p} \underbrace{(a_{j_1 j_1 r_2}, \dot{x}_{j_2})}_{(m_{r_2}^{(2)})^{\text{th}}\text{order XX}}$$
$$\{a_{j_1 j_1 r_1} \rightleftharpoons a_{j_1 j_1 r_2} | m_{r_1}^{(1)} = m_{r_2}^{(2)}; r_1, r_2 \in \{1, 2, \ldots, p\}, r_1 \neq r_2\};$$

and

$$\begin{aligned}&a_{j_1 j_1 r_1} = a_{j_1 j_1 1} - \varepsilon_{r_1}, a_{j_1 j_1 r_2} = a_{j_1 j_1 1} + \varepsilon_{r_2},\\ &(a_{j_1 j_1 r_1} = a_{j_1 j_1 1} + \varepsilon_{r_1}, a_{j_1 j_1 r_2} = a_{j_1 j_1 1} - \varepsilon_{r_2});\\ &\varepsilon_{r_1}, \varepsilon_{r_2} > 0; \ r_i = 1, 2, \ldots, p \ (i = 1, 2).\end{aligned}$$

- Two 1-dimensional flows of $(a_{j_1 j_2 s_1}, \dot{x}_{j_2})$ and $(a_{j_1 j_2 s_1}, \dot{x}_{j_2})$ switching at point $(a_{j_1 j_2 1}, \dot{x}_{j_2})$ with the same order singularity are given through

$$\bigcup_{r_1=1}^{q} \underbrace{(a_{j_1 j_2 s_1}, \dot{x}_{j_2})}_{(n_{s_1}^{(1)})^{\text{th}}\text{order YY}} \rightleftharpoons \underbrace{(a_{j_1 j_2 1}, \dot{x}_{j_2})}_{(2n_1+1)^{\text{th}} \text{ order YY}} \rightleftharpoons \bigcup_{s_2=1}^{q} \underbrace{(a_{j_1 j_2 s_2}, \dot{x}_{j_2})}_{(n_{s_2}^{(1)})^{\text{th}}\text{order YY}}$$
$$\{a_{j_1 j_2 s_1} \rightleftharpoons a_{j_1 j_2 s_2} | n_{s_1}^{(1)} = n_{s_2}^{(2)}; s_1, s_2 \in \{1, 2, \ldots, q\}, s_1 \neq s_2\};$$

and

$$a_{j_1j_2s_1} = a_{j_1j_2 1} - \varepsilon_{s_1}, a_{j_1j_2s_2} = a_{j_1j_2 1} + \varepsilon_{s_2},$$
$$(a_{j_1j_2s_1} = a_{j_1j_2 1} + \varepsilon_{s_1}, a_{j_1j_2s_2} = a_{j_1j_2 1} - \varepsilon_{s_2});$$
$$\varepsilon_{s_1}, \varepsilon_{s_2} > 0;\ s_i = 1, 2, \ldots, q\,(i = 1, 2).$$

- Two 1-dimensional flows of $(a_{j_1j_1r_1}, a_{j_1j_2s_1})$ and $(a_{j_1j_1r_2}, a_{j_1j_2s_2})$ switching at point $(a_{j_1j_1 1}, a_{j_1j_2 1})$ with the same order singularity are given through

$$\bigcup_{\alpha_1=1}^{p}\bigcup_{\beta_1=1}^{q} \underbrace{(a_{j_1j_1\alpha_1}, a_{j_1j_2\beta_1})}_{([(m^{(1)}_{\alpha_1}),(n^{(1)}_{\beta_1})]\text{XX,xF})\text{-flow}} \rightleftharpoons \underbrace{([(2m_1+1),(2\text{n}_1+1)]\text{XX, xF})}_{[(2m_1+1),(2\text{n}_1+1)]\text{-XX}}$$
$$\rightleftharpoons \bigcup_{\alpha_2=1}^{p}\bigcup_{\beta_2=1}^{q} \underbrace{(a_{j_1j_1\alpha_2}, a_{j_1j_2\beta_2})}_{([(m^{(2)}_{\alpha_2}),(n^{(2)}_{\beta_2})]\text{XX,xF})\text{-flow}}$$

with

$$a_{j_1j_1\alpha_1} = a_{j_1j_1 1} - \varepsilon_{\alpha_1}, a_{j_1j_1\alpha_2} = a_{j_1j_1 1} + \varepsilon_{\alpha_2},$$
$$(a_{j_1j_1\alpha_1} = a_{j_1j_1 1} + \varepsilon_{\alpha_1}, a_{j_1j_1\alpha_2} = a_{j_1j_1 1} - \varepsilon_{\alpha_2}),$$
$$\varepsilon_{\alpha_1}, \varepsilon_{\alpha_2} > 0,\ \alpha_i = 1, 2, \ldots, p\,(i = 1, 2);$$
$$a_{j_1j_2\beta_1} = a_{j_1j_2 1} - \varepsilon_{\beta_1}, a_{j_1j_2\beta_2} = a_{j_1j_2 1} + \varepsilon_{\beta_2},$$
$$(a_{j_1j_2\beta_1} = a_{j_1j_2 1} + \varepsilon_{\beta_1}, a_{j_1j_2\beta_2} = a_{j_1j_2 1} - \varepsilon_{\beta_2}),$$
$$\varepsilon_{s_1}, \varepsilon_{s_2} > 0,\ s_i = 1, 2, \ldots, q\,(i = 1, 2).$$

In summary, from the analysis in case (I_1)–(I_3), the 1-dimensional flow of $(x^*_{j_1}, x^*_{j_1}) = (a_{j_1j_1 1}, a_{j_2j_1 1})$ for $m = 2m_1 + 1$ and $n = 2n_1 + 1$ has the bifurcation properties as stated in (vi_{1a})–(vi_{1d}) through Eqs. (1.83–1.98).

(vi) For $m = 2m_1$ and $n = 2n_1 + 1$, the equilibrium of $(x^*_{j_1}, x^*_{j_2}) = (a_{j_1j_1 1}, a_{j_1j_2 1})$ for the x_{j_1}-direction has the following bifurcation properties with

$$2m_1 - \sum_{\alpha_1=1}^{p_1} m_{\alpha_1} = 2\sum_{\rho_1=1}^{l_1} m_{\rho_1}, 2n_1 + 1 - \sum_{\beta_1=1}^{q_1} n_{\beta_1} = 2\sum_{\sigma_1=1}^{l_2} n_{\sigma_1};$$
$$\sum_{\alpha_2=1}^{p_2} m_{\alpha_2} = 2m_1, \sum_{\beta_2=1}^{q_2} n_{\beta_2} = 2n_1 + 1;$$
$$\sum_{\alpha_i=1}^{p} m_{\alpha_i} = 2m_1, \sum_{\beta_i=1}^{q} n_{\beta_i} = 2n_1 + 1, \text{ for } i = 1, 2; \tag{2.1}$$

l_1, l_2-quadratic polynomials are without real roots and ρ_1^{th} and σ_1^{th}-quadratic polynomials are with powers m_{ρ_1} and n_{σ_1},respectively.

The bifurcation process is discussed through differential equations as follows. There are four bifurcations (II$_1$)–(II$_3$).

(II$_1$) Consider a dynamical system having a single equilibrium (A$_1$) as

$$\begin{aligned}\dot{x}_{j_1} &= a_{j_1j_10}\prod_{r=1}^{m_1}[(x_{j_1}-a_{j_1j_1}^{(r)})^2+\Delta_{j_1j_1}^{(r)}]\\&\quad\times(x_{j_1}-b_{j_2j_11})\prod_{s=1}^{n_1}[(x_{j_1}-a_{j_2j_1}^{(s)})^2+\Delta_{j_2j_1}^{(s)}],\\ \dot{x}_{j_2} &= a_{j_2j_10};\end{aligned}$$

where $a_{j_2j_11}=b_{j_2j_11}$. Once

$$\begin{aligned}\Delta_{j_1j_1}^{(r)} &= \Delta_{j_1j_1}=0\,(r=1,2,\ldots,m_1),\\ \Delta_{j_1j_2}^{(s)} &= \Delta_{j_1j_2}=0\,(r=1,2,\ldots,n_1),\end{aligned}$$

if

$$\begin{aligned}a_{j_1j_11} &= a_{j_1j_1}^{(r)}\;(r=1,2,\ldots,m_1),\\ a_{j_1j_21} &= b_{j_1j_21}=a_{j_1j_2}^{(s)}\;(s=1,2,\ldots,n_1)\end{aligned}$$

the foregoing differential equations becomes a dynamical system with a singular equilibrium (B$_1$) as

$$\dot{x}_{j_1}=a_{j_1j_10}(x_{j_1}-a_{j_1j_11})^{2m_1}(x_{j_2}-a_{j_1j_21})^{2n_1+1},\;\dot{x}_{j_2}=a_{j_2j_20}.$$

If

$$\begin{aligned}\Delta_{j_1j_1}^{(r)} &= \Delta_{j_1j_1}+\delta_r\,(r=1,2,\ldots,m_1),\;\delta_r>0;\\ \Delta_{j_1j_2}^{(s)} &= \Delta_{j_1j_2}+\delta_s\,(s=1,2,\ldots,n_1),\;\delta_s>0;\end{aligned}$$

then

$$\begin{aligned}&a_{j_1j_11}^{(r)},a_{j_1j_12}^{(r)}=a_{j_1j_11}\pm\varepsilon_r,\;(r=1,2,\ldots,m_1),\\&\{a_{j_1j_11},a_{j_1j_12},\ldots,a_{j_1j_1(2m_1)}\}=\text{sort}\{\,a_{j_1j_11}^{(r)},a_{j_1j_12}^{(r)}|r=1,2,\ldots,m_1\}\;;\\&a_{j_1j_21}^{(s)},a_{j_1j_22}^{(s)}=a_{j_1j_21}\pm\varepsilon_s,\;(s=1,2,\ldots,n_1),\\&\{a_{j_1j_21},a_{j_1j_22},\ldots,a_{j_1j_2(2n_1+1)}\}=\text{sort}\{\,b_{j_1j_21},a_{j_1j_21}^{(s)},a_{j_1j_22}^{(s)}|s=1,2,\ldots,n_1\}\,.\end{aligned}$$

Thus, the differential equation becomes a dynamical system with a non-singular equilibrium network (C_1) as

$$\dot{x}_{j_1} = a_{j_1 j_1 0} \prod_{\alpha=1}^{2m_1} (x_{j_1} - a_{j_1 j_1 \alpha}) \prod_{\beta=1}^{2n_1+1} (x_{j_2} - a_{j_1 j_2 \beta}),\ \dot{x}_{j_2} = a_{j_2 j_2 0}.$$

- From the above bifurcation process analysis, at least, only m_1-parameter variations in the x_{j_1}-direction are engaged in such a bifurcation from a 1-dimensional flow to the 1-dimensional flow array of $1 \times (2m_1)$ through the $(2m_1)^{\text{th}}$-upper-saddle and lower-saddle bifurcations in the x_{j_1}-direction. *The higher-order upper-saddle and lower-saddle flow bifurcations are developed.*

  Thus, the 1-dimensional flow appearing or vanishing bifurcation route is as follows.

$$\underbrace{(\dot{x}_{j_1}, \dot{x}_{j_2})}_{\text{(xF,xF)-flow}} \rightleftharpoons \underbrace{(a_{j_1 j_1 1}, \dot{x}_{j_2})}_{(2m_1+1)^{\text{th}}\text{order-XX}} \rightleftharpoons \bigcup_{r=1}^{2m_1} \underbrace{(a_{j_1 j_1 r}, \dot{x}_{j_2})}_{\text{XX}}.$$

- From the above bifurcation process analysis, at least, only $(n_1 + 1)$-parameter variations in the x_{j_1}-direction are engaged in such a bifurcation from a 1-dimensional flow to the 1-dimensional flow array of $(2n_1 + 1) \times 1$ through the $(2n_1 + 1)^{\text{th}}$-up-parabola and down-parabola bifurcations in the x_{j_1}-direction. *The higher-order up-parabola and lower-parabola flow bifurcations are developed.*

  Thus, the 1-dimensional flow appearing or vanishing bifurcation route is as follows.

$$\underbrace{(a_{j_1 j_2 1}, \dot{x}_{j_2})}_{\text{YY}} \rightleftharpoons \underbrace{(a_{j_1 j_2 1}, \dot{x}_{j_2})}_{(2n_1+1)^{\text{th}}\text{order-YY}} \rightleftharpoons \bigcup_{\beta=1}^{2n_1+1} \underbrace{(a_{j_1 j_2 \beta}, \dot{x}_{j_2})}_{\text{YY}}.$$

- From the above bifurcation process analysis, $(m_1 + n_1 + 1)$-parameter variations in the two directions are engaged in such a bifurcation from a parabola-flow to a $(2m_1) \times (2n_1 + 1)$ network of hyperbolic and hyperbolic-secant flows through the $(2m_1)^{\text{th}}$-upper-saddle and lower-saddle bifurcations in the x_{j_1}-direction, and the $(2n_1 + 1)^{\text{th}}$-up-parabola and down-parabola flows bifurcations in the x_{j_1}-direction. *The higher-order hyperbolic to hyperbolic-recant flow bifurcations are developed.*

  Thus, the 1-dimensional flow appearing or vanishing bifurcation route is as follows

$$\underbrace{(a_{j_1 j_1 1}, a_{j_1 j_2 1})}_{\text{(XX,xF)-flow}} \rightleftharpoons \underbrace{[(2m_1), (2n_1 + 1)]\text{XX}, xF)}_{[(2m_1),(2n_1+1)]\text{-XX}}$$

$$\rightleftharpoons \bigcup_{\alpha=1}^{2m_1} \bigcup_{\beta=1}^{2n_1+1} \underbrace{(a_{j_1j_1\alpha}, a_{j_1j_2\beta})}_{\text{(YY,xF)-flow}}.$$

(II_2) From Case (II_1), consider a dynamical system having singular equilibriums (A_2) as

$$\begin{aligned}\dot{x}_{j_1} &= a_{j_1j_10} \prod_{r_1}^{p_1} (x_{j_1} - b_{j_1j_1r_1})^{m_{r_1}} \prod_{r=1}^{l_1} [(x_{j_1} - a_{j_1j_1}^{(r)})^2 + \Delta_{j_1j_1}^{(r)}]^{m_r} \\ &\quad \times \prod_{r_1}^{q_1} (x_{j_2} - b_{j_1j_2r_1})^{m_{r_1}} \prod_{s=1}^{l_2} [(x_{j_2} - a_{j_1j_2}^{(s)})^2 + \Delta_{j_1j_2}^{(s)}]^{m_s}, \\ \dot{x}_{j_2} &= a_{j_2j_20};\end{aligned}$$

where

$$\begin{aligned}&2m_1 - \sum_{\alpha_1=1}^{p_1} m_{\alpha_1} = 2\sum_{\rho_1=1}^{l_1} m_{\rho_1}, \; 2n_1 + 1 - \sum_{\beta_1=1}^{q_1} n_{\beta_1} = 2\sum_{\sigma_1=1}^{l_2} n_{\sigma_1}; \\ &\sum_{\alpha_2=1}^{p_2} m_{\alpha_2} = 2m_1, \; \sum_{\beta_2=1}^{q_2} n_{\beta_2} = 2n_1 + 1;\end{aligned}$$

l_1, l_2-quadratic polynomials are without real roots and ρ_1^{th} and σ_1^{th}-quadratic polynomials are with powers m_{ρ_1} and n_{σ_1},respectively.

Once

$$\begin{aligned}\Delta_{j_1j_1}^{(r)} &= 0\,(r = 1, 2, \ldots, l_1), \\ \Delta_{j_1j_2}^{(s)} &= 0\,(s = 1, 2, \ldots, l_2);\end{aligned}$$

if

$$\begin{aligned}a_{j_1j_11} &\equiv a_{j_1j_1r_1} = a_{j_1j_1}^{(r)}\,(r_1 = 1, 2, \ldots, p_1;\, r = 1, 2, \ldots, l_1), \\ a_{j_1j_21} &\equiv a_{j_1j_2r_1} = a_{j_1j_2}^{(s)}\,(s_1 = 1, 2, \ldots, q_1;\, s = 1, 2, \ldots, l_2)\end{aligned}$$

the foregoing differential equations becomes a dynamical system with a 1-dimensional flow (B_2) as

$$\dot{x}_{j_1} = a_{j_1j_10}(x_{j_1} - a_{j_1j_11})^{2m_1}(x_{j_2} - a_{j_1j_21})^{2n_1+1}, \; \dot{x}_{j_2} = a_{j_2j_10}.$$

If

$$\Delta_{j_1j_1}^{(r)} = \Delta_{j_1j_1} + \delta_r\ (r = 1, 2, \ldots, l_1),\ \delta_r > 0;$$
$$\Delta_{j_1j_2}^{(s)} = \Delta_{j_1j_2} + \delta_s\ (s = 1, 2, \ldots, l_2),\ \delta_s > 0;$$

then

$$\begin{aligned}
&a_{j_1j_11}^{(r)}, a_{j_1j_12}^{(r)} = a_{j_1j_11} \pm \varepsilon_r,\ (r = 1, 2, \ldots, l_1),\\
&\{a_{j_1j_11}, a_{j_1j_12}, \ldots, a_{j_1j_1p_2}\}\\
&= \text{sort}\,\{a_{j_1j_11}, a_{j_1j_12}, \ldots, a_{j_1j_1p_1}; a_{j_1j_11}^{(r)}, a_{j_1j_12}^{(r)} | r = 1, 2, \ldots, l_1\};\\
&a_{j_1j_21}^{(s)}, a_{j_1j_22}^{(s)} = a_{j_1j_21} \pm \varepsilon_s,\ (s = 1, 2, \ldots, l_2),\\
&\{a_{j_1j_21}, a_{j_1j_22}, \ldots, a_{j_1j_2q_2}\}\\
&= \text{sort}\,\{a_{j_1j_21}, a_{j_1j_22}, \ldots, a_{j_1j_2q_1}; a_{j_2j_11}^{(s)}, a_{j_2j_12}^{(s)} | s = 1, 2, \ldots, l_2\}.
\end{aligned}$$

Thus, the differential equation becomes a dynamical system with singular 1-dimensional flows (C_2) as

$$\begin{aligned}
\dot{x}_{j_1} &= a_{j_1j_10} \prod_{\alpha_2}^{p_2} (x_{j_1} - a_{j_1j_1\alpha_2})^{m_{\alpha_2}} \prod_{\beta_2}^{q_2} (x_{j_2} - a_{j_1j_1\beta_2})^{n_{\beta_2}},\\
\dot{x}_{j_2} &= a_{j_2j_10}
\end{aligned}$$

where

$$\Sigma_{\alpha_2=1}^{p_2} m_{\alpha_2}^{(2)} = 2m_1,\ \Sigma_{\beta_2=1}^{q_2} n_{\beta_2}^{(2)} = 2n_1 + 1.$$

- From the above bifurcation process analysis, only p_2-parameter variations in the x_{j_1}-direction from a p_1 to p_2 1-dimensional flow array through the $(2m_1)^{\text{th}}$-source and sink flow bifurcations in the x_{j_1}-direction.

  Thus, the 1-dimensional flow appearing or vanishing bifurcation route from a p_1 to p_2 1-dimensional flow array is expressed as

$$\bigcup_{r_1=1}^{p_1} \underbrace{(a_{j_1j_1r_1}, \dot{x}_{j_2})}_{(\text{m}_{r_1}^{(1)})^{\text{th}}\text{order XX}} \rightleftharpoons \underbrace{(a_{j_1j_11}, \dot{x}_{j_2})}_{(2m_1)^{\text{th}}\text{ order XX}} \rightleftharpoons \bigcup_{r_2=1}^{p_2} \underbrace{(a_{j_1j_1r_2}, \dot{x}_{j_2})}_{(\text{m}_{r_2}^{(2)})^{\text{th}} order\text{ XX}}.$$

- From the above bifurcation process analysis, only q_2-parameter variations in the x_{j_1}-direction from a q_1 to q_2 1-dimensional flow array through the $(2n_1 + 1)^{\text{th}}$-up-parabola and down-parabola flow bifurcations in the x_{j_1}-direction.

  Thus, the 1-dimensional flow appearing or vanishing bifurcation route from a q_1 to q_2 1-dimensional flow array is expressed as

$$\bigcup_{\beta_1=1}^{q_1}\underbrace{(a_{j_2j_1\beta_1},\dot{x}_{j_2})}_{(n_{\beta_1}^{(1)})^{\text{th}}\text{order YY}} \rightleftharpoons \underbrace{(a_{j_1j_2 1},\dot{x}_{j_2})}_{(2n_1+1)^{\text{th}}\text{ order YY}} \rightleftharpoons \bigcup_{s_2=1}^{q_2}\underbrace{(a_{j_1j_2\beta_2},\dot{x}_{j_2})}_{(n_{\beta_2}^{(2)})^{\text{th}}\text{order YY}}.$$

- From the above bifurcation process analysis, only q_2+p_2-parameter variations in the x_{j_1}-direction from a singular $q_1\times p_1$ to $q_2\times p_2$ network of 1-dimensional flows through the $(2m_1)^{\text{th}}$-upper-saddle and lower-saddle flows bifurcations in the x_{j_1}-direction, and the $(2n_1+1)^{\text{th}}$-up-parabola and down-parabola flows bifurcations in the x_{j_1}-direction.

  Thus, the 1-dimensional flow appearing or vanishing bifurcation route is as follows

$$\bigcup_{\alpha_1=1}^{p_1}\bigcup_{\beta_1=1}^{q_1}\underbrace{(a_{j_1j_1\alpha_1},a_{j_1j_2\beta_1})}_{([(m_{r_1}),(n_{s_1})]\text{XX,xF)-flow}} \rightleftharpoons \underbrace{[(2m_1),(2n_1+1)]\text{XX, xF)}}_{[(2m_1),(2n_1+1)]\text{-XX}}$$
$$\rightleftharpoons \bigcup_{\alpha_2=1}^{p_2}\bigcup_{\beta_2=1}^{q_2}\underbrace{(a_{j_1j_1\alpha_2},a_{j_1j_2\beta_2})}_{([(m_{r_1}),(n_{s_1})]\text{XX,xF)-flow}}.$$

(II_3) Consider two dynamical systems with the same 1-dimensional flows with locations switched (A_3,C_3) for $i=1,2$ as

$$\dot{x}_{j_1}=a_{j_1j_1 0}\prod_{r_i}^{p}(x_{j_1}-a_{j_1j_1r_i})^{m_{r_i}^{(i)}}\prod_{s_i}^{q}(x_{j_1}-a_{j_2j_1s_i})^{n_{s_i}^{(i)}},$$
$$\dot{x}_{j_2}=a_{j_2j_2 0};$$

where

$$\Sigma_{r_i=1}^{p}m_{r_i}^{(i)}=2m_1,\ \Sigma_{r_i=1}^{q}n_{r_i}^{(i)}=2n_1+1.$$

Consider a dynamical system as a singular 1-dimensional flow (B_3) as

$$\dot{x}_{j_1}=a_{j_1j_1 0}(x_{j_1}-a_{j_1j_1 1})^{2m_1}(x_{j_1}-a_{j_1j_2 1})^{2n_1+1},\ \dot{x}_{j_2}=a_{j_2j_2 0}.$$

For $r_i=1,2,\ldots,p$ and $s_i=1,2,\ldots,q$ ($i=1,2$), define a function as

$$\Delta_{j_1j_1}^{(r_1r_2)}=(a_{j_1j_1r_1}-a_{j_1j_1r_2})^2\ (r_1,r_2\in\{1,2,\ldots,p\},r_1\neq r_2),$$
$$\Delta_{j_1j_2}^{(s_1s_2)}=(a_{j_1j_2s_1}-a_{j_1j_2s_2})^2\ (s_1,s_2\in\{1,2,\ldots,q\},s_1\neq s_2).$$

For $r_i=1,2,\ldots,p$ and $s_i=1,2,\ldots,q$ ($i=1,2$), if

$$\Delta_{j_1j_1}^{(r_1r_2)}=(a_{j_1j_1r_1}-a_{j_1j_1r_2})^2=0\text{ and }\Delta_{j_1j_2}^{(s_1s_2)}=(a_{j_1j_2s_1}-a_{j_1j_2s_2})^2=0,$$

- Two 1-dimensional flows of $(a_{j_1j_1r_1},\dot{x}_{j_2})$ and $\big(a_{j_1j_1r_2},\dot{x}_{j_2}\big)$ switching at point $\big(a_{j_1j_1r_1},\dot{x}_{j_2}\big)$ with the same order singularity are given through

$$\bigcup_{r_1=1}^{p}\underbrace{(a_{j_1j_1r_1},\dot{x}_{j_2})}_{(m_{r_1}^{(1)})^{\text{th}}\text{order XX}}\rightleftharpoons\underbrace{(a_{j_1j_11},\dot{x}_{j_2})}_{(2m_1)^{\text{th}}\text{ order XX}}\rightleftharpoons\bigcup_{r_2=1}^{p}\underbrace{(a_{j_1j_1r_2},\dot{x}_{j_2})}_{(m_{r_2}^{(2)})^{\text{th}}\text{order XX}}$$
$$\big\{a_{j_1j_1r_1}\rightleftharpoons a_{j_1j_1r_2}|m_{r_1}^{(1)}=m_{r_2}^{(2)};r_1,r_2\in\big\{1,2,\ldots,p\big\},r_1\neq r_2\big\};$$

and

$$\begin{aligned}&a_{j_1j_1r_1}=a_{j_1j_11}-\varepsilon_{r_1},a_{j_1j_1r_2}=a_{j_1j_11}+\varepsilon_{r_2},\\&(a_{j_1j_1r_1}=a_{j_1j_11}+\varepsilon_{r_1},a_{j_1j_1r_2}=a_{j_1j_11}-\varepsilon_{r_2});\\&\varepsilon_{r_1},\varepsilon_{r_2}>0;\ r_i=1,2,\ldots,p\,(i=1,2).\end{aligned}$$

- Two 1-dimensional flows of $(a_{j_1j_2s_1},\dot{x}_{j_2})$ and $(a_{j_1j_2s_1},\dot{x}_{j_2})$ switching at point $(a_{j_1j_21},\dot{x}_{j_2})$ with the same order singularity are given through

$$\bigcup_{r_1=1}^{q}\underbrace{(a_{j_1j_2s_1},\dot{x}_{j_2})}_{(n_{s_1}^{(1)})^{\text{th}}\text{order YY}}\rightleftharpoons\underbrace{(a_{j_1j_21},\dot{x}_{j_2})}_{(2n_1+1)^{\text{th}}\text{ order YY}}\rightleftharpoons\bigcup_{s_2=1}^{q}\underbrace{(a_{j_1j_2s_2},\dot{x}_{j_2})}_{(n_{s_2}^{(1)})^{\text{th}}\text{order YY}}$$
$$\big\{a_{j_1j_2s_1}\rightleftharpoons a_{j_1j_2s_2}|n_{s_1}^{(1)}=n_{s_2}^{(2)};s_1,s_2\in\big\{1,2,\ldots,q\big\},s_1\neq s_2\big\};$$

and

$$\begin{aligned}&a_{j_1j_2s_1}=a_{j_1j_21}-\varepsilon_{s_1},a_{j_1j_2s_2}=a_{j_1j_21}+\varepsilon_{s_2},\\&(a_{j_1j_2s_1}=a_{j_1j_21}+\varepsilon_{s_1},a_{j_1j_2s_2}=a_{j_1j_21}-\varepsilon_{s_2});\\&\varepsilon_{s_1},\varepsilon_{s_2}>0;\ s_i=1,2,\ldots,q\,(i=1,2).\end{aligned}$$

- Two 1-dimensional flows of $(a_{j_1j_1r_1},a_{j_1j_2s_1})$ and $(a_{j_1j_1r_2},a_{j_1j_2s_2})$ switching at point $(a_{j_1j_11},a_{j_1j_21})$ with the same order singularity are given through

$$\begin{aligned}&\bigcup_{\alpha_1=1}^{p}\bigcup_{\beta_1=1}^{q}\underbrace{(a_{j_1j_1\alpha_1},a_{j_1j_2\beta_1})}_{([(m_{\alpha_1}^{(1)}),(n_{\beta_1}^{(1)})]\text{XX,xF)-flow}}\rightleftharpoons\underbrace{([(2m_1),(2\text{n}_1+1)]\text{XX, xF})}_{[(2m_1),(2\text{n}_1+1)]\text{-XX}}\\&\rightleftharpoons\bigcup_{\alpha_2=1}^{p}\bigcup_{\beta_2=1}^{q}\underbrace{(a_{j_1j_1\alpha_2},a_{j_1j_2\beta_2})}_{([(m_{\alpha_2}^{(2)}),(n_{\beta_2}^{(2)})]\text{XX,xF)-flow}}\end{aligned}$$

with

$$
\begin{aligned}
&a_{j_1j_1\alpha_1} = a_{j_1j_1 1} - \varepsilon_{\alpha_1}, a_{j_1j_1\alpha_2} = a_{j_1j_1 1} + \varepsilon_{\alpha_2},\\
&(a_{j_1j_1\alpha_1} = a_{j_1j_1 1} + \varepsilon_{\alpha_1}, a_{j_1j_1\alpha_2} = a_{j_1j_1 1} - \varepsilon_{\alpha_2}),\\
&\varepsilon_{\alpha_1}, \varepsilon_{\alpha_2} > 0,\ \alpha_i = 1, 2, \ldots, p\,(i = 1, 2);\\
&a_{j_1j_2\beta_1} = a_{j_1j_2 1} - \varepsilon_{\beta_1}, a_{j_1j_2\beta_2} = a_{j_1j_2 1} + \varepsilon_{\beta_2},\\
&(a_{j_1j_2\beta_1} = a_{j_1j_2 1} + \varepsilon_{\beta_1}, a_{j_1j_2\beta_2} = a_{j_1j_2 1} - \varepsilon_{\beta_2}),\\
&\varepsilon_{s_1}, \varepsilon_{s_2} > 0,\ s_i = 1, 2, \ldots, q\,(i = 1, 2).
\end{aligned}
$$

In summary, from cases (II_1)–(II_3), the 1-dimensional flows of $(x_{j_1}^*, x_{j_2}^*) = (a_{j_1j_1 1}, a_{j_1j_2 1})$ for $m = 2m_1$ *and* $n = 2n_1 + 1$ has the bifurcation properties as stated in (vi_{1a})–(vi_{1d}) through Eqs. (1.100)–(1.115).

(vii) For $m = 2m_1 + 1$ and $n = 2n_1$, the equilibrium of $(x_{j_1}^*, x_{j_2}^*) = (a_{j_1j_1 1}, a_{j_1j_2 1})$ for the x_{j_1}-direction has the following bifurcation properties with

$$
\begin{aligned}
&2m_1 + 1 - \sum_{\alpha_1=1}^{p_1} m_{\alpha_1} = 2\sum_{\rho_1=1}^{l_1} m_{\rho_1}, 2n_1 - \sum_{\beta_1=1}^{q_1} n_{\beta_1} = 2\sum_{\sigma_1=1}^{l_2} n_{\sigma_1};\\
&\sum_{\alpha_2=1}^{p_2} m_{\alpha_2} = 2m_1 + 1, \sum_{\beta_2=1}^{q_2} n_{\beta_2} = 2n_1;\\
&\sum_{\alpha_i=1}^{p} m_{\alpha_i} = 2m_1 + 1, \sum_{\beta_i=1}^{q} n_{\beta_i} = 2n_1,\ \text{for } i = 1, 2;
\end{aligned}
$$

l_1, l_2-quadratic polynomials are without real roots and ρ_1^{th} and σ_1^{th}-quadratic polynomials are with powers m_{ρ_1} and n_{σ_1},respectively.

The bifurcation process is discussed through differential equations as follows. There are four bifurcations (III_1)–(III_3).

(III_1) Consider a dynamical system having a single equilibrium (A_1) as

$$
\begin{aligned}
&\dot{x}_{j_1} = a_{j_1j_1 0}(x_{j_1} - b_{j_1j_1 1})\prod_{r=1}^{m_1}[(x_{j_1} - a_{j_1j_1}^{(r)})^2 + \Delta_{j_1j_1}^{(r)}]\prod_{s=1}^{n_1}[(x_{j_1} - a_{j_1j_2}^{(s)})^2 + \Delta_{j_1j_2}^{(s)}],\\
&\dot{x}_{j_2} = a_{j_2j_2 0};
\end{aligned}
$$

where $a_{j_1j_1 1} = b_{j_1j_1 1}$. Once

$$
\begin{aligned}
&\Delta_{j_1j_1}^{(r)} = \Delta_{j_1j_1} = 0(r = 1, 2, \ldots, m_1),\\
&\Delta_{j_1j_2}^{(s)} = \Delta_{j_1j_2} = 0(r = 1, 2, \ldots, n_1)
\end{aligned}
$$

if

$$
\begin{aligned}
a_{j_1j_11} &= b_{j_1j_11} = a^{(r)}_{j_1j_1}\ (r = 1, 2, \ldots, m_1),\\
a_{j_1j_21} &= b_{j_1j_21} = a^{(s)}_{j_1j_2}\ (s = 1, 2, \ldots, n_1)
\end{aligned}
$$

the foregoing differential equations becomes a dynamical system with a singular equilibrium (B_1) as

$$
\dot{x}_{j_1} = a_{j_1j_10}(x_{j_1} - a_{j_1j_11})^{2m_1+1}(x_{j_2} - a_{j_1j_21})^{2n_1},\ \dot{x}_{j_2} = a_{j_2j_20}.
$$

If

$$
\begin{aligned}
\Delta^{(r)}_{j_1j_1} &= \Delta_{j_1j_1} + \delta_r\ (r = 1, 2, \ldots, m_1),\ \delta_r > 0;\\
\Delta^{(s)}_{j_1j_2} &= \Delta_{j_1j_2} + \delta_s\ (s = 1, 2, \ldots, n_1),\ \delta_s > 0;
\end{aligned}
$$

then

$$
\begin{aligned}
&a^{(r)}_{j_1j_11}, a^{(r)}_{j_1j_12} = a_{j_1j_11} \pm \varepsilon_r,\ (r = 1, 2, \ldots, m_1),\\
&\left\{a_{j_1j_11}, a_{j_1j_12}, \ldots, a_{j_1j_1(2m_1+1)}\right\} = \text{sort}\left\{b_{j_1j_11}, a^{(r)}_{j_1j_11}, a^{(r)}_{j_1j_12} | r = 1, 2, \ldots, m_1\right\};\\
&a^{(s)}_{j_1j_21}, a^{(s)}_{j_1j_22} = a_{j_1j_21} \pm \varepsilon_s,\ (s = 1, 2, \ldots, n_1),\\
&\left\{a_{j_1j_21}, a_{j_1j_22}, \ldots, a_{j_1j_2(2n_1+1)}\right\} = \text{sort}\left\{b_{j_1j_21}, a^{(s)}_{j_1j_21}, a^{(s)}_{j_1j_22} | s = 1, 2, \ldots, n_1\right\}.
\end{aligned}
$$

Thus, the differential equation becomes a dynamical system with a non-singular equilibrium network (C_1) as

$$
\dot{x}_{j_1} = a_{j_1j_10} \prod_{\alpha=1}^{2m_1+1} (x_{j_1} - a_{j_1j_1\alpha}) \prod_{\beta=1}^{2n_1} (x_{j_2} - a_{j_1j_2\beta}),\ \dot{x}_{j_2} = a_{j_2j_20}.
$$

- From the above bifurcation process analysis, at least, only (m_1+1)-parameter variations in the x_{j_1}-direction are engaged in such a bifurcation from a 1-dimensional flow to the 1-dimensional flow array of $1 \times (2m_1 + 1)$ through the $(2m_1 + 1)^{\text{th}}$-source and sink bifurcations in the x_{j_1}-direction. *The higher-order source and sink flow bifurcations are developed.*

  Thus, the 1-dimensional flow appearing or vanishing bifurcation route is as follows.

$$\underbrace{(a_{j_1j_1 1}, \dot{x}_{j_2})}_{\text{XX}} \rightleftharpoons \underbrace{(a_{j_1j_1 1}, \dot{x}_{j_2})}_{(2m_1+1)^{\text{th}}\text{order-XX}} \rightleftharpoons \bigcup_{\alpha=1}^{2m_1+1} \underbrace{(a_{j_1j_1\alpha}, \dot{x}_{j_2})}_{\text{XX}}.$$

- From the above bifurcation process analysis, at least, only n_1-parameter variations in the x_{j_1}-direction are engaged in such a bifurcation from a 1-dimensional flow to the 1-dimensional flow array of $(2n_1) \times 1$ through the $(2n_1)^{\text{th}}$-increasing and inflection flows bifurcations in the x_{j_1}-direction. *The higher-order decreasing to increasing inflection flow bifurcations are developed.*

  Thus, the 1-dimensional flow appearing or vanishing bifurcation route is as follows.

$$\underbrace{(\dot{x}_{j_2}, \dot{x}_{j_2})}_{\text{(xF,xF)-flow}} \rightleftharpoons \underbrace{(a_{j_1j_2 1}, \dot{x}_{j_2})}_{(2n_1)^{\text{th}}\text{order-YY}} \rightleftharpoons \bigcup_{\beta=1}^{2n_1} \underbrace{(a_{j_1j_2\beta}, \dot{x}_{j_2})}_{\text{YY}}.$$

- From the above bifurcation process analysis, $(m_1 + n_1 + 1)$-parameter variations in the two directions are engaged in such a bifurcation from a parabola-flow to a $(2m_1 + 1) \times (2n_1)$ network of hyperbolic and hyperbolic-secant flows through the $(2m_1 + 1)^{th}$-source and sink bifurcations in the x_{j_1}-direction, and the $(2n_1)^{\text{th}}$-increasing-inflection and decreasing-inflection flows bifurcations in the x_{j_1}-direction. *The higher-order decreasing-to-increasing inflection source and sink flows bifurcations are developed.*

  Thus, the 1-dimensional source and sink flow appearing or vanishing bifurcation route is as follows.

$$\begin{aligned}&\underbrace{(a_{j_1j_1 1}, \dot{x}_{j_2})}_{\text{(XX,xF)-flow}} \rightleftharpoons \underbrace{[(2m_1+1), (2n_1)]\text{XX}, xF)}_{[(2m_1+1),(2n_1)]\text{-XX}}\\ &\rightleftharpoons \bigcup_{\alpha=1}^{2m_1+1}\bigcup_{\beta=1}^{2n_1} \underbrace{(a_{j_1j_1\alpha}, a_{j_1j_2\beta})}_{\text{(YY,xF)-flow}}.\end{aligned}$$

(III_2) From Case (III_1), consider a dynamical system having singular equilibriums (A_2) as

$$\begin{aligned}\dot{x}_{j_1} =& a_{j_1j_1 0} \prod_{r_1=1}^{p_1} (x_{j_1} - b_{j_1j_1r_1})^{m_{r_1}} \prod_{r=1}^{l_1} [(x_{j_1} - a_{j_1j_1}^{(r)})^2 + \Delta_{j_1j_1}^{(r)}]^{m_r}\\ &\times \prod_{s_1=1}^{q_1} (x_{j_2} - b_{j_1j_2s_1})^{n_{s_1}} \prod_{s=1}^{l_2} [(x_{j_2} - a_{j_1j_2}^{(s)})^2 + \Delta_{j_1j_2}^{(s)}]^{n_s},\\ \dot{x}_{j_2} =& a_{j_2j_1 0};\end{aligned}$$

where

$$2m_1+1-\sum_{\alpha_1=1}^{p_1} m_{\alpha_1}=2\sum_{\rho_1=1}^{l_1} m_{\rho_1},\ 2n_1-\sum_{\beta_1=1}^{q_1} n_{\beta_1}=2\sum_{\sigma_1=1}^{l_2} n_{\sigma_1};$$
$$\sum_{\alpha_2=1}^{p_2} m_{\alpha_2}=2m_1+1,\ \sum_{\beta_2=1}^{q_2} n_{\beta_2}=2n_1;$$

l_1, l_2-quadratic polynomials are without real roots and ρ_1^{th} and σ_1^{th}-quadratic polynomials are with powers m_{ρ_1} and n_{σ_1},respectively.

Once

$$\Delta_{j_1j_1}^{(r)}=0\,(r=1,2,\ldots,l_1),$$
$$\Delta_{j_1j_2}^{(s)}=0\,(s=1,2,\ldots,l_2);$$

if

$$a_{j_1j_11}\equiv a_{j_1j_1r_1}=a_{j_1j_1}^{(r)}\ (r_1=1,2,\ldots,p_1;\ r=1,2,\ldots,l_1),$$
$$a_{j_1j_21}\equiv a_{j_1j_2r_1}=a_{j_1j_2}^{(s)}\ (s_1=1,2,\ldots,q_1;\ s=1,2,\ldots,l_2)$$

the foregoing differential equations becomes a dynamical system with a 1-dimensional flow (B_2) as

$$\dot{x}_{j_1}=a_{j_1j_10}(x_{j_1}-a_{j_1j_11})^{2m_1+1}(x_{j_2}-a_{j_1j_21})^{2n_1},\ \dot{x}_{j_2}=a_{j_2j_20}.$$

If

$$\Delta_{j_1j_1}^{(r)}=\Delta_{j_1j_1}+\delta_r\,(r=1,2,\ldots,l_1),\ \delta_r>0;$$
$$\Delta_{j_1j_2}^{(s)}=\Delta_{j_1j_2}+\delta_s\,(s=1,2,\ldots,l_2),\ \delta_s>0;$$

then

$$\begin{aligned}
&a_{j_1j_11}^{(r)}, a_{j_1j_12}^{(r)}=a_{j_1j_11}\pm\varepsilon_r,\ (r=1,2,\ldots,l_1),\\
&\{a_{j_1j_11}, a_{j_1j_12},\ldots,a_{j_1j_1p_2}\}\\
&=\text{sort}\{\ a_{j_1j_11}, a_{j_1j_12},\ldots,a_{j_1j_1p_1};\ a_{j_1j_11}^{(r)}, a_{j_1j_12}^{(r)}|r=1,2,\ldots,l_1\}\ ;\\
&a_{j_1j_21}^{(s)}, a_{j_1j_22}^{(s)}=a_{j_1j_21}\pm\varepsilon_s,\ (s=1,2,\ldots,l_2),\\
&\{a_{j_1j_21}, a_{j_1j_22},\ldots,a_{j_1j_2q_2}\}\\
&=\text{sort}\{\ a_{j_1j_21}, a_{j_1j_22},\ldots,a_{j_1j_2q_1};\ a_{j_1j_21}^{(s)}, a_{j_1j_22}^{(s)}|s=1,2,\ldots,l_2\}\ .
\end{aligned}$$

Thus, the differential equation becomes a dynamical system with singular 1-dimensional flows (C_2) as

$$\dot{x}_{j_1} = a_{j_1 j_1 0} \prod_{\alpha_2=1}^{p_2} (x_{j_1} - a_{j_1 j_1 \alpha_2})^{m_{\alpha 2}} \prod_{\beta_2=1}^{q_2} (x_{j_2} - a_{j_1 j_1 \beta_2})^{n_{\beta_2}},$$

$$\dot{x}_{j_2} = a_{j_2 j_2 0}$$

where

$$\Sigma_{\alpha_2=1}^{p_2} m_{\alpha_2}^{(2)} = 2m_1 + 1,\ \Sigma_{\beta_2=1}^{q_2} n_{\beta_2}^{(2)} = 2n_1.$$

- From the above bifurcation process analysis, only p_2-parameter variations in the x_{j_1}-direction from a p_1 to p_2 1-dimensional flow array through the $(2m_1 + 1)^{\text{th}}$-source and sink flow bifurcations in the x_{j_1}-direction.

  Thus, the 1-dimensional flow appearing or vanishing bifurcation route from a p_1 to p_2 1-dimensional flow array is expressed as

$$\bigcup_{r_1=1}^{p_1} \underbrace{(a_{j_1 j_1 r_1}, \dot{x}_{j_2})}_{(\mathrm{m}_{r_1}^{(1)})^{\text{th}}\text{order XX}} \rightleftharpoons \underbrace{(a_{j_1 j_1 1}, \dot{x}_{j_2})}_{(2m_1+1)^{\text{th}}\text{ order XX}} \rightleftharpoons \bigcup_{r_2=1}^{p_2} \underbrace{(a_{j_1 j_1 r_2}, \dot{x}_{j_2})}_{(\mathrm{m}_{r_2}^{(2)})^{\text{th}}\text{order XX}}.$$

- From the above bifurcation process analysis, only q_2-parameter variations in the x_{j_1}-direction from a q_1 to q_2 1-dimensional flow array through the $(2n_1)^{\text{th}}$-increasing and decreasing inflection flow bifurcations in the x_{j_1}-direction.

  Thus, the 1-dimensional flow appearing or vanishing bifurcation route from a q_1 to q_2 1-dimensional flow array is expressed as

$$\bigcup_{\beta_1=1}^{q_1} \underbrace{(a_{j_1 j_2 \beta_1}, \dot{x}_{j_2})}_{(n_{\beta_1}^{(1)})^{\text{th}}\text{order YY}} \rightleftharpoons \underbrace{(a_{j_1 j_2 1}, \dot{x}_{j_2})}_{(2n_1)^{\text{th}}\text{ order YY}} \rightleftharpoons \bigcup_{s_2=1}^{q_2} \underbrace{(a_{j_1 j_2 \beta_2}, \dot{x}_{j_2})}_{(n_{\beta_2}^{(2)})^{\text{th}}\text{order YY}}.$$

- From the above bifurcation process analysis, only $q_2 + p_2$-parameter variations in the x_{j_1}-direction from a $q_1 \times p_1$ to $q_2 \times p_2$ network of 1-dimensional flows through the $(2m_1 + 1)^{\text{th}}$-source and sink flows bifurcations in the x_{j_1}-direction, and the $(2n_1)^{\text{th}}$-increasing and decreasing inflection flows bifurcations in the x_{j_1}-direction.

  Thus, the 1-dimensional flow appearing or vanishing bifurcation route is as follows.

$$\bigcup_{\alpha_1=1}^{p_1} \bigcup_{\beta_1=1}^{q_1} \underbrace{(a_{j_1 j_1 \alpha_1}, a_{j_1 j_2 \beta_1})}_{([(m_{r_1}),(n_{s_1})]\text{XX,xF})\text{-flow}} \rightleftharpoons \underbrace{[(2m_1{+}1),\ (2n_1)]\text{XX, xF})}_{[(2m_1+1),(2n_1)]-\text{XX}}$$
$$\rightleftharpoons \bigcup_{\alpha_2=1}^{p_2} \bigcup_{\beta_2=1}^{q_2} \underbrace{(a_{j_1 j_1 \alpha_2}, a_{j_1 j_2 \beta_2})}_{([(m_{r_1}),(n_{s_1})]\text{XX,xF})\text{-flow}}.$$

(III_3) Consider two dynamical systems with the same 1-dimensional flows with locations switched (A_3,C_3) for $i = 1, 2$ as

$$\dot{x}_{j_1} = a_{j_1 j_1 0} \prod_{r_i=1}^{p} (x_{j_1} - a_{j_1 j_1 r_i})^{m_{r_i}^{(i)}} \prod_{s_i=1}^{q} (x_{j_1} - a_{j_2 j_1 s_i})^{n_{s_i}^{(i)}},$$

$$\dot{x}_{j_2} = a_{j_2 j_1 0};$$

where

$$\Sigma_{r_i=1}^{p} m_{r_i}^{(i)} = 2m_1 + 1, \ \Sigma_{r_i=1}^{q} n_{r_i}^{(i)} = 2n_1.$$

Consider a dynamical system as a singular 1-dimensional flow (B_3) as

$$\dot{x}_{j_1} = a_{j_1 j_1 0}(x_{j_1} - a_{j_1 j_1 1})^{2m_1+1}(x_{j_1} - a_{j_2 j_1 1})^{2n_1}, \ \dot{x}_{j_2} = a_{j_2 j_2 0}.$$

For $r_i = 1, 2, \ldots, p$ and $s_i = 1, 2, \ldots, q$ $(i = 1, 2)$, define a function as

$$\Delta_{j_1 j_1}^{(r_1 r_2)} = (a_{j_1 j_1 r_1} - a_{j_1 j_1 r_2})^2 \ (r_1, r_2 \in \{1, 2, \ldots, p\}, r_1 \neq r_2),$$
$$\Delta_{j_1 j_2}^{(s_1 s_2)} = (a_{j_1 j_2 s_1} - a_{j_1 j_2 s_2})^2 \ (s_1, s_2 \in \{1, 2, \ldots, q\}, s_1 \neq s_2).$$

For $r_i = 1, 2, \ldots, p$ and $s_i = 1, 2, \ldots, q$ $(i = 1, 2)$, if

$$\Delta_{j_1 j_1}^{(r_1 r_2)} = (a_{j_1 j_1 r_1} - a_{j_1 j_1 r_2})^2 = 0 \text{ and } \Delta_{j_1 j_2}^{(s_1 s_2)} = (a_{j_1 j_2 s_1} - a_{j_1 j_2 s_2})^2 = 0,$$

- Two 1-dimensional flows of $(a_{j_1 j_1 r_1}, \dot{x}_{j_2})$ and $(a_{j_1 j_1 r_2}, \dot{x}_{j_2})$ switching at point $(a_{j_1 j_1 r_1}, \dot{x}_{j_2})$ with the same order singularity are given through

$$\bigcup_{r_1=1}^{p} \underbrace{(a_{j_1 j_1 r_1}, \dot{x}_{j_2})}_{(m_{r_1}^{(1)})^{\text{th}} \text{order XX}} \rightleftharpoons \underbrace{(a_{j_1 j_1 1}, \dot{x}_{j_2})}_{(2m_1+1)^{\text{th}} \text{ order XX}} \rightleftharpoons \bigcup_{r_2=1}^{p} \underbrace{(a_{j_1 j_1 r_2}, \dot{x}_{j_2})}_{(m_{r_2}^{(2)})^{\text{th}} \text{order XX}}$$
$$\{a_{j_1 j_1 r_1} \rightleftharpoons a_{j_1 j_1 r_2} | m_{r_1}^{(1)} = m_{r_2}^{(2)}; r_1, r_2 \in \{1, 2, \ldots, p\}, r_1 \neq r_2\};$$

and

$$a_{j_1 j_1 r_1} = a_{j_1 j_1 1} - \varepsilon_{r_1}, a_{j_1 j_1 r_2} = a_{j_1 j_1 1} + \varepsilon_{r_2},$$
$$(a_{j_1 j_1 r_1} = a_{j_1 j_1 1} + \varepsilon_{r_1}, a_{j_1 j_1 r_2} = a_{j_1 j_1 1} - \varepsilon_{r_2});$$
$$\varepsilon_{r_1}, \varepsilon_{r_2} > 0; \ r_i = 1, 2, \ldots, p \ (i = 1, 2).$$

- Two 1-dimensional flows of $(a_{j_1j_2s_1}, \dot{x}_{j_2})$ and $(a_{j_1j_2s_1}, \dot{x}_{j_2})$ switching at point $(a_{j_1j_2 1}, \dot{x}_{j_2})$ with the same order singularity are given through

$$\bigcup_{r_1=1}^{q} \underbrace{(a_{j_1j_2s_1}, \dot{x}_{j_2})}_{(n_{s_1}^{(1)})^{\text{th}}\text{order YY}} \rightleftharpoons \underbrace{(a_{j_1j_2 1}, \dot{x}_{j_2})}_{(2n_1)^{\text{th}}\text{ order YY}} \rightleftharpoons \bigcup_{s_2=1}^{q} \underbrace{(a_{j_1j_2s_2}, \dot{x}_{j_2})}_{(n_{s_2}^{(1)})^{\text{th}}\text{order YY}}$$
$$\left\{a_{j_1j_2s_1} \rightleftharpoons a_{j_1j_2s_2} | n_{s_1}^{(1)} = n_{s_2}^{(2)}; s_1, s_2 \in \{1, 2, \ldots, q\}, s_1 \neq s_2\right\};$$

and

$$\begin{aligned} &a_{j_1j_2s_1} = a_{j_1j_2 1} - \varepsilon_{s_1}, a_{j_1j_2s_2} = a_{j_1j_2 1} + \varepsilon_{s_2}, \\ &(a_{j_1j_2s_1} = a_{j_1j_2 1} + \varepsilon_{s_1}, a_{j_1j_2s_2} = a_{j_1j_2 1} - \varepsilon_{s_2}); \\ &\varepsilon_{s_1}, \varepsilon_{s_2} > 0;\ s_i = 1, 2, \ldots, q\,(i = 1, 2). \end{aligned}$$

- Two 1-dimensional flows of $(a_{j_1j_1r_1}, a_{j_1j_2s_1})$ and $(a_{j_1j_1r_2}, a_{j_1j_2s_2})$ switching at point $(a_{j_1j_1 1}, a_{j_1j_2 1})$ with the same order singularity are given through

$$\bigcup_{\alpha_1=1}^{p} \bigcup_{\beta_1=1}^{q} \underbrace{(a_{j_1j_1\alpha_1}, a_{j_1j_2\beta_1})}_{([(m_{\alpha_1}^{(1)}),(n_{\beta_1}^{(1)})]\text{XX,xF)-flow}} \rightleftharpoons \underbrace{([(2m_1+1),(2n_1)]\text{XX, xF})}_{[(2m_1+1),(2n_1)]\text{-XX}}$$
$$\rightleftharpoons \bigcup_{\alpha_2=1}^{p} \bigcup_{\beta_2=1}^{q} \underbrace{(a_{j_1j_1\alpha_2}, a_{j_1j_2\beta_2})}_{([(m_{\alpha_2}^{(2)}),(n_{\beta_2}^{(2)})]\text{XX,xF)-flow}}$$

with

$$\begin{aligned} &a_{j_1j_1\alpha_1} = a_{j_1j_1 1} - \varepsilon_{\alpha_1}, a_{j_1j_1\alpha_2} = a_{j_1j_1 1} + \varepsilon_{\alpha_2}, \\ &(a_{j_1j_1\alpha_1} = a_{j_1j_1 1} + \varepsilon_{\alpha_1}, a_{j_1j_1\alpha_2} = a_{j_1j_1 1} - \varepsilon_{\alpha_2}), \\ &\varepsilon_{\alpha_1}, \varepsilon_{\alpha_2} > 0,\ \alpha_i = 1, 2, \ldots, p\,(i = 1, 2); \\ &a_{j_1j_2\beta_1} = a_{j_1j_2 1} - \varepsilon_{\beta_1}, a_{j_1j_2\beta_2} = a_{j_1j_2 1} + \varepsilon_{\beta_2}, \\ &(a_{j_1j_2\beta_1} = a_{j_1j_2 1} + \varepsilon_{\beta_1}, a_{j_1j_2\beta_2} = a_{j_1j_2 1} - \varepsilon_{\beta_2}), \\ &\varepsilon_{s_1}, \varepsilon_{s_2} > 0,\ s_i = 1, 2, \ldots, q\,(i = 1, 2). \end{aligned}$$

In summary, from cases (III_1)–(III_3), the 1-dimensional flow of $(x_{j_1}^*, x_{j_2}^*) = (a_{j_1j_1 1}, a_{j_1j_2 1})$ for $m = 2m_1 + 1$ and $n = 2n_1$ has the bifurcation properties as stated in (vii_{1a})–(vii_{1d}) through Eqs. (1.117–1.132).

(viii) For $m = 2m_1$ and $n = 2n_1$, the equilibrium of $(x_{j_1}^*, x_{j_2}^*) = (a_{j_1j_1 1}, a_{j_1j_2 1})$ for the x_{j_1}-direction has the following bifurcation properties with

$$
\begin{aligned}
&2m_1 - \sum_{\alpha_1=1}^{p_1} m_{\alpha_1} = 2\sum_{\rho_1=1}^{l_1} m_{\rho_1},\ 2n_1 - \sum_{\beta_1=1}^{q_1} n_{\beta_1} = 2\sum_{\sigma_1=1}^{l_2} n_{\sigma_1};\\
&\sum_{\alpha_2=1}^{p_2} m_{\alpha_2} = 2m_1,\ \sum_{\beta_2=1}^{q_2} n_{\beta_2} = 2n_1;\\
&\sum_{\alpha_i=1}^{p} m_{\alpha_i} = 2m_1,\ \sum_{\beta_i=1}^{q} n_{\beta_i} = 2n_1,\ \text{for } i = 1, 2;
\end{aligned}
\tag{2.2}
$$

l_1, l_2-quadratic polynomials are without real roots and ρ_1^{th} and σ_1^{th}-quadratic polynomials are with powers m_{ρ_1} and n_{σ_1}, respectively.

The bifurcation process is discussed through differential equations as follows. There are four bifurcations (IV_1)–(IV_3).

(IV_1) Consider a dynamical system having a single equilibrium (A_1) as

$$
\begin{aligned}
\dot{x}_{j_1} &= a_{j_1j_10}\prod_{r=1}^{m_1}[(x_{j_1} - a_{j_1j_1}^{(r)})^2 + \Delta_{j_1j_1}^{(r)}] \times \prod_{s=1}^{n_1}[(x_{j_1} - a_{j_2j_1}^{(s)})^2 + \Delta_{j_2j_1}^{(s)}],\\
\dot{x}_{j_2} &= a_{j_2j_20}.
\end{aligned}
$$

Once

$$
\begin{aligned}
\Delta_{j_1j_1}^{(r)} &= \Delta_{j_1j_1} + 0(r = 1, 2, \ldots, m_1),\\
\Delta_{j_1j_2}^{(s)} &= \Delta_{j_1j_2} + 0(r = 1, 2, \ldots, n_2),
\end{aligned}
$$

if

$$
\begin{aligned}
a_{j_1j_11} &= b_{j_1j_11} = a_{j_1j_1}^{(r)}\ (r = 1, 2, \ldots, m_1),\\
a_{j_1j_21} &= b_{j_1j_21} = a_{j_1j_2}^{(s)}\ (s = 1, 2, \ldots, n_1)
\end{aligned}
$$

the foregoing differential equations becomes a dynamical system with a singular equilibrium (B_1) as

$$
\dot{x}_{j_1} = a_{j_1j_10}(x_{j_1} - a_{j_1j_11})^{2m_1}(x_{j_2} - a_{j_1j_21})^{2n_1},\ \dot{x}_{j_2} = a_{j_2j_20}.
$$

If

$$
\begin{aligned}
\Delta_{j_1j_1}^{(r)} &= \Delta_{j_1j_1} + \delta_r\ (r = 1, 2, \ldots, m_1),\ \delta_r > 0;\\
\Delta_{j_2j_1}^{(s)} &= \Delta_{j_2j_1} + \delta_s\ (s = 1, 2, \ldots, n_2),\ \delta_s > 0;
\end{aligned}
$$

then

$$\begin{aligned}
&a^{(r)}_{j_1j_11}, a^{(r)}_{j_1j_12} = a_{j_1j_11} \pm \varepsilon_r,\ (r = 1, 2, \ldots, m_1),\\
&\{a_{j_1j_11}, a_{j_1j_12}, \ldots, a_{j_1j_1(2m_1)}\} = \text{sort}\ \{a^{(r)}_{j_1j_11}, a^{(r)}_{j_1j_12} | r = 1, 2, \ldots, m_1\};\\
&a^{(s)}_{j_1j_21}, a^{(s)}_{j_1j_22} = a_{j_1j_21} \pm \varepsilon_s,\ (s = 1, 2, \ldots, n_1),\\
&\{a_{j_1j_21}, a_{j_1j_22}, \ldots, a_{j_1j_2(2n_1)}\} = \text{sort}\ \{a^{(s)}_{j_1j_21}, a^{(s)}_{j_1j_22} | s = 1, 2, \ldots, n_1\}.
\end{aligned}$$

Thus, the differential equation becomes a dynamical system with a non-singular equilibrium network (C_1) as

$$\dot{x}_{j_1} = a_{j_1j_10} \prod_{\alpha=1}^{2m_1} (x_{j_1} - a_{j_1j_1\alpha}) \prod_{\beta=1}^{2n_1} (x_{j_2} - a_{j_1j_2\beta}),\ \dot{x}_{j_2} = a_{j_2j_20}.$$

- From the above bifurcation process analysis, at least, only m_1-parameter variations in the x_{j_1}-direction are engaged in such a bifurcation from a 1-dimensional flow to the 1-dimensional flow array of $1 \times (2m_1)$ through the $(2m_1)^{\text{th}}$-upper-saddle and lower-saddle flow bifurcations in the x_{j_1}-direction. *The higher-order upper-saddle and lower-saddle flow bifurcations are developed.*

  Thus, the 1-dimensional flow appearing or vanishing bifurcation route is as follows.

$$\underbrace{(a_{j_1j_11}, \dot{x}_{j_2})}_{\text{XX}} \rightleftharpoons \underbrace{(a_{j_1j_11}, \dot{x}_{j_2})}_{(2m_1+1)^{\text{th}}\text{order-XX}} \rightleftharpoons \bigcup_{\alpha=1}^{2m_1} \underbrace{(a_{j_1j_1\alpha}, \dot{x}_{j_2})}_{\text{XX}}.$$

- From the above bifurcation process analysis, at least, only n_1-parameter variations in the x_{j_1}-direction are engaged in such a bifurcation from a 1-dimensional flow to the 1-dimensional flow array of $(2n_1) \times 1$ through the $(2n_1)^{\text{th}}$-increasing and inflection flows bifurcations in the x_{j_1}-direction. *The higher-order decreasing to increasing inflection flow bifurcations are developed.*

  Thus, the 1-dimensional flow appearing or vanishing bifurcation route is as follows.

$$\underbrace{(\dot{x}_{j_2}, \dot{x}_{j_2})}_{\text{(xF,xF)-flow}} \rightleftharpoons \underbrace{(a_{j_1j_21}, \dot{x}_{j_2})}_{(2n_1)^{\text{th}}\text{order-YY}} \rightleftharpoons \bigcup_{\beta=1}^{2n_1} \underbrace{(a_{j_1j_2\beta}, \dot{x}_{j_2})}_{\text{YY}}.$$

- From the above bifurcation process analysis, $(m_1 + n_1)$-parameter variations in the two directions are engaged in such a bifurcation from a parabola-flow to a $(2m_1) \times (2n_1)$-network of hyperbolic and hyperbolic-secant flows through the $(2m_1)^{\text{th}}$-upper-saddle and lower-saddle bifurcations in the x_{j_1}-direction, and the $(2n_1)^{\text{th}}$-increasing-inflection and decreasing-inflection flows bifurcations in the x_{j_1}-direction. *The higher-order decreasing-to-increasing inflection upper-saddle and lower-saddle flows bifurcations are developed.*

  Thus, the 1-dimensional source and sink flow appearing or vanishing bifurcation route is as follows.

$$\underbrace{(\dot{x}_{j_1}, \dot{x}_{j_2})}_{\text{(xF,xF)-flow}} \rightleftharpoons \underbrace{[(2m_1), (2n_1)]\text{XX}, xF)}_{[(2m_1),(2n_1)]\text{-}XX} \rightleftharpoons \underbrace{\bigcup_{\alpha=1}^{2m_1} \bigcup_{\beta=1}^{2n_1} (a_{j_1 j_1 \alpha}, a_{j_1 j_2 \beta})}_{\text{(YY,xF)-flow}}.$$

(IV_2): From Case (IV_1), consider a dynamical system having singular equilibriums (A_2) as

$$\begin{aligned}
\dot{x}_{j_1} =& a_{j_1 j_1 0} \prod_{r_1}^{p_1} (x_{j_1} - b_{j_1 j_1 r_1})^{m_{r_1}} \prod_{r=1}^{l_1} [(x_{j_1} - a_{j_1 j_1}^{(r)})^2 + \Delta_{j_1 j_1}^{(r)}]^{m_r} \\
& \times \prod_{r_1}^{q_1} (x_{j_2} - b_{j_1 j_2 r_1})^{m_{r_1}} \prod_{s=1}^{l_2} [(x_{j_2} - a_{j_1 j_2}^{(s)})^2 + \Delta_{j_1 j_2}^{(s)}]^{m_s}, \\
\dot{x}_{j_2} =& a_{j_2 j_2 0};
\end{aligned}$$

where

$$\begin{aligned}
& 2m_1 + 1 - \sum_{\alpha_1=1}^{p_1} m_{\alpha_1} = 2 \sum_{\rho_1=1}^{l_1} m_{\rho_1}, \; 2n_1 + 1 - \sum_{\beta_1=1}^{q_1} n_{\beta_1} = 2 \sum_{\sigma_1=1}^{l_2} n_{\sigma_1}; \\
& \sum_{\alpha_2=1}^{p_2} m_{\alpha_2} = 2m_1 + 1, \; \sum_{\beta_2=1}^{q_2} n_{\beta_2} = 2n_1 + 1;
\end{aligned}$$

l_1, l_2-quadratic polynomials are without real roots and ρ_1^{th} and σ_1^{th}-quadratic polynomials are with powers m_{ρ_1} and n_{σ_1},respectively.

Once

$$\begin{aligned}
\Delta_{j_1 j_1}^{(r)} &= 0 \; (r = 1, 2, \ldots, l_1), \\
\Delta_{j_1 j_2}^{(s)} &= 0 \; (s = 1, 2, \ldots, l_2);
\end{aligned}$$

if

$$a_{j_1j_11} \equiv a_{j_1j_1r_1} = a^{(r)}_{j_1j_1}\ (r_1 = 1, 2, \ldots, p_1; r = 1, 2, \ldots, l_1),$$
$$a_{j_1j_21} \equiv a_{j_1j_2r_1} = a^{(s)}_{j_1j_2}\ (s_1 = 1, 2, \ldots, q_1; s = 1, 2, \ldots, l_2),$$

the foregoing differential equations becomes a dynamical system with a 1-dimensional flow (B_2) as

$$\dot{x}_{j_1} = a_{j_1j_10}(x_{j_1} - a_{j_1j_11})^{2m_1}(x_{j_2} - a_{j_1j_21})^{2n_1},\ \dot{x}_{j_2} = a_{j_2j_20}.$$

If

$$\Delta^{(r)}_{j_1j_1} = \Delta_{j_1j_1} + \delta_r\ (r = 1, 2, \ldots, l_1),\ \delta_r > 0;$$
$$\Delta^{(s)}_{j_1j_2} = \Delta_{j_1j_2} + \delta_s\ (s = 1, 2, \ldots, l_2),\ \delta_s > 0;$$

then

$$\begin{aligned}
&a^{(r)}_{j_1j_12}, a^{(r)}_{j_1j_12} = a_{j_1j_11} \pm \varepsilon_r,\ (r = 1, 2, \ldots, l_1),\\
&\{a_{j_1j_11}, a_{j_1j_12}, \ldots, a_{j_1j_1p_2}\}\\
&= \text{sort}\{a_{j_1j_11}, a_{j_1j_12}, \ldots, a_{j_1j_1p_1}; a^{(r)}_{j_1j_11}, a^{(r)}_{j_1j_12} | r = 1, 2, \ldots, l_1\};\\
&a^{(s)}_{j_1j_21}, a^{(s)}_{j_1j_22} = a_{j_1j_21} \pm \varepsilon_s,\ (s = 1, 2, \ldots, l_2),\\
&\{a_{j_1j_21}, a_{j_1j_22}, \ldots, a_{j_1j_2q_2}\}\\
&= \text{sort}\{a_{j_1j_21}, a_{j_1j_22}, \ldots, a_{j_1j_2q_1}; a^{(s)}_{j_1j_21}, a^{(s)}_{j_1j_22} | s = 1, 2, \ldots, l_2\}.
\end{aligned}$$

Thus, the differential equation becomes a dynamical system with singular 1-dimensional flows (C_2) as

$$\dot{x}_{j_1} = a_{j_1j_10}\prod_{\alpha_2=1}^{p_2}(x_{j_1} - a_{j_1j_1\alpha_2})^{m_{\alpha_2}}\prod_{\beta_2=1}^{q_2}(x_{j_2} - a_{j_1j_1\beta_2})^{n_{\beta_2}},$$
$$\dot{x}_{j_2} = a_{j_2j_20}$$

where

$$\Sigma^{p_2}_{\alpha_2=1}m^{(2)}_{\alpha_2} = 2m_1,\ \Sigma^{q_2}_{\beta_2=1}n^{(2)}_{\beta_2} = 2n_1.$$

- From the above bifurcation process analysis, only p_2-parameter variations in the x_{j_1}-direction from a p_1 to p_2 1-dimensional flow array through the $(2m_1)^{\text{th}}$-upper-saddle and lower-saddle flow bifurcations in the x_{j_1}-direction.

  Thus, the 1-dimensional flow appearing or vanishing bifurcation route from a p_1 to p_2 1-dimensional flow array is expressed as

$$\bigcup_{r_1=1}^{p_1} \underbrace{(a_{j_1 j_1 r_1}, \dot{x}_{j_2})}_{(\mathrm{m}_{r_1}^{(1)})^{\text{th}}\text{order XX}} \rightleftharpoons \underbrace{(a_{j_1 j_1 1}, \dot{x}_{j_2})}_{(2m_1)^{\text{th}}\text{ order XX}} \rightleftharpoons \bigcup_{r_2=1}^{p_2} \underbrace{(a_{j_1 j_1 r_2}, \dot{x}_{j_2})}_{(\mathrm{m}_{r_2}^{(2)})^{\text{th}}\text{order XX}}.$$

- From the above bifurcation process analysis, only q_2-parameter variations in the x_{j_1}-direction from a q_1 to q_2 1-dimensional flow array through the $(2n_1)^{\text{th}}$-increasing and decreasing inflection flow bifurcations in the x_{j_1}-direction.

  Thus, the 1-dimensional flow appearing or vanishing bifurcation route from a q_1 to q_2 1-dimensional flow array is expressed as

$$\bigcup_{\beta_1=1}^{q_1} \underbrace{(a_{j_1 j_2 \beta_1}, \dot{x}_{j_2})}_{(n_{\beta_1}^{(1)})^{\text{th}}\text{order YY}} \rightleftharpoons \underbrace{(a_{j_1 j_2 1}, \dot{x}_{j_2})}_{(2n_1)^{\text{th}}\text{ order YY}} \rightleftharpoons \bigcup_{s_2=1}^{q_2} \underbrace{(a_{j_1 j_2 \beta_2}, \dot{x}_{j_2})}_{(n_{\beta_2}^{(2)})^{\text{th}}\text{order YY}}.$$

- From the above bifurcation process analysis, only $q_2 + p_2$-parameter variations in the x_{j_1}-direction from a $q_1 \times p_1$ to $q_2 \times p_2$ network of 1-dimensional flows through the $(2m_1)^{\text{th}}$-upper-saddle and lower-saddle flows bifurcations in the x_{j_1}-direction, and the $(2n_1)^{\text{th}}$-increasing and decreasing inflection flows bifurcations in the x_{j_1}-direction.

  Thus, the 1-dimensional flow appearing or vanishing bifurcation route is as follows

$$\begin{aligned} &\bigcup_{\alpha_1=1}^{p_1} \bigcup_{\beta_1=1}^{q_1} \underbrace{(a_{j_1 j_1 \alpha_1}, a_{j_1 j_2 \beta_1})}_{([(m_{r_1}),(n_{s_1})]\text{XX,xF})\text{-flow}} \rightleftharpoons \underbrace{[(2m_1), (2n_1)]\text{XX, xF})}_{[(2m_1),(2n_1)]\text{-XX}} \\ &\rightleftharpoons \bigcup_{\alpha_2=1}^{p_2} \bigcup_{\beta_2=1}^{q_2} \underbrace{(a_{j_1 j_1 \alpha_2}, a_{j_1 j_2 \beta_2})}_{([(m_{r_1}),(n_{s_1})]\text{XX,xF})\text{-flow}}. \end{aligned}$$

(IV$_3$) Consider two dynamical systems with the same 1-dimensional flows with locations switched (A$_3$,C$_3$) for $i = 1, 2$ as

$$\begin{aligned} \dot{x}_{j_1} &= a_{j_1 j_1 0} \prod_{r_i=1}^{p} (x_{j_1} - a_{j_1 j_1 r_i})^{m_{r_i}^{(i)}} \prod_{s_i=1}^{q} (x_{j_1} - a_{j_2 j_1 s_i})^{n_{s_i}^{(i)}}, \\ \dot{x}_{j_2} &= a_{j_2 j_2 0}; \end{aligned}$$

where

$$\Sigma_{r_i=1}^{p} m_{r_i}^{(i)} = 2m_1,\ \Sigma_{r_i=1}^{q} n_{r_i}^{(i)} = 2n_1.$$

Consider a dynamical system as a singular 1-dimensional flow (B_3) as

$$\dot{x}_{j_1} = a_{j_1 j_1 0}(x_{j_1} - a_{j_1 j_1 1})^{2m_1}(x_{j_2} - a_{j_1 j_2 1})^{2n_1},\ \dot{x}_{j_2} = a_{j_2 j_2 0}.$$

For $r_i = 1, 2, \ldots, p$ and $s_i = 1, 2, \ldots, q$ $(i = 1, 2)$, define a function as

$$\Delta_{j_1 j_1}^{(r_1 r_2)} = (a_{j_1 j_1 r_1} - a_{j_1 j_1 r_2})^2\ (r_1, r_2 \in \{1, 2, \ldots, p\}, r_1 \neq r_2),$$
$$\Delta_{j_1 j_2}^{(s_1 s_2)} = (a_{j_1 j_2 s_1} - a_{j_1 j_2 s_2})^2\ (s_1, s_2 \in \{1, 2, \ldots, q\}, s_1 \neq s_2).$$

For $r_i = 1, 2, \ldots, p$ and $s_i = 1, 2, \ldots, q$ $(i = 1, 2)$, if

$$\Delta_{j_1 j_1}^{(r_1 r_2)} = (a_{j_1 j_1 r_1} - a_{j_1 j_1 r_2})^2 = 0 \text{ and } \Delta_{j_1 j_2}^{(s_1 s_2)} = (a_{j_1 j_2 s_1} - a_{j_1 j_2 s_2})^2 = 0,$$

- Two 1-dimensional flows of $(a_{j_1 j_2 r_1}, \dot{x}_{j_2})$ and $(a_{j_1 j_1 r_2}, \dot{x}_{j_2})$ switching at point $(a_{j_1 j_1 1}, \dot{x}_{j_2})$ with the same order singularity are given through

$$\bigcup_{r_1=1}^{p} \underbrace{(a_{j_1 j_1 r_1}, \dot{x}_{j_2})}_{(m_{r_1}^{(1)})^{\text{th}} \text{order XX}} \rightleftharpoons \underbrace{(a_{j_1 j_1 1}, \dot{x}_{j_2})}_{(2m_1)^{\text{th}} \text{ order XX}} \rightleftharpoons \bigcup_{r_2=1}^{p} \underbrace{(a_{j_1 j_1 r_2}, \dot{x}_{j_2})}_{(m_{r_2}^{(2)})^{\text{th}} \text{order XX}}$$
$$\{a_{j_1 j_1 r_1} \rightleftharpoons a_{j_1 j_1 r_2} | m_{r_1}^{(1)} = m_{r_2}^{(2)}; r_1, r_2 \in \{1, 2, \ldots, p\}, r_1 \neq r_2\};$$

and

$$a_{j_1 j_1 r_1} = a_{j_1 j_1 1} - \varepsilon_{r_1}, a_{j_1 j_1 r_2} = a_{j_1 j_1 1} + \varepsilon_{r_2},$$
$$(a_{j_1 j_1 r_1} = a_{j_1 j_1 1} + \varepsilon_{r_1}, a_{j_1 j_1 r_2} = a_{j_1 j_1 1} - \varepsilon_{r_2});$$
$$\varepsilon_{r_1}, \varepsilon_{r_2} > 0;\ r_i = 1, 2, \ldots, p\ (i = 1, 2).$$

- Two 1-dimensional flows of $(a_{j_1 j_2 s_1}, \dot{x}_{j_2})$ and $(a_{j_1 j_2 s_1}, \dot{x}_{j_2})$ switching at point $(a_{j_1 j_2 1}, \dot{x}_{j_2})$ with the same order singularity are given through

$$\bigcup_{r_1=1}^{q} \underbrace{(a_{j_1 j_2 s_1}, \dot{x}_{j_2})}_{(n_{s_1}^{(1)})^{\text{th}} \text{order YY}} \rightleftharpoons \underbrace{(a_{j_1 j_2 1}, \dot{x}_{j_2})}_{(2n_1)^{\text{th}} \text{ order YY}} \rightleftharpoons \bigcup_{s_2=1}^{q} \underbrace{(a_{j_1 j_2 s_2}, \dot{x}_{j_2})}_{(n_{s_2}^{(1)})^{\text{th}} \text{order YY}}$$
$$\left\{a_{j_1 j_2 s_1} \rightleftharpoons a_{j_1 j_2 s_2} | n_{s_1}^{(1)} = n_{s_2}^{(2)}; s_1, s_2 \in \left\{1, 2, \ldots, q\right\}, s_1 \neq s_2\right\};$$

and

$$\begin{aligned}
&a_{j_1j_2s_1} = a_{j_1j_2 1} - \varepsilon_{s_1}, a_{j_1j_2s_2} = a_{j_1j_2 1} + \varepsilon_{s_2},\\
&(a_{j_1j_2s_1} = a_{j_1j_2 1} + \varepsilon_{s_1}, a_{j_1j_2s_2} = a_{j_1j_2 1} - \varepsilon_{s_2});\\
&\varepsilon_{s_1}, \varepsilon_{s_2} > 0;\ s_i = 1, 2, \ldots, q\,(i = 1, 2).
\end{aligned}$$

- Two 1-dimensional flows of $(a_{j_1j_1r_1}, a_{j_1j_2s_1})$ and $(a_{j_1j_1r_2}, a_{j_1j_2s_2})$ switching at point $(a_{j_1j_1 1}, a_{j_1j_2 1})$ with the same order singularity are given through

$$\begin{aligned}
&\bigcup_{\alpha_1=1}^{p}\bigcup_{\beta_1=1}^{q} \underbrace{(a_{j_1j_1\alpha_1}, a_{j_1j_2\beta_1})}_{([(m^{(1)}_{\alpha_1}),(n^{(1)}_{\beta_1})]\text{XX,xF})\text{-}flow} \rightleftharpoons \underbrace{([(2m_1),(2\mathrm{n}_1)]\text{XX, xF})}_{[(2m_1),(2\mathrm{n}_1)]\text{-XX}}\\
&\rightleftharpoons \bigcup_{\alpha_2=1}^{p}\bigcup_{\beta_2=1}^{q} \underbrace{(a_{j_1j_1\alpha_2}, a_{j_1j_2\beta_2})}_{([(m^{(2)}_{\alpha_2}),(n^{(2)}_{\beta_2})]\text{XX,xF})\text{-}flow},
\end{aligned}$$

with

$$\begin{aligned}
&a_{j_1j_1\alpha_1} = a_{j_1j_1 1} - \varepsilon_{\alpha_1}, a_{j_1j_1\alpha_2} = a_{j_1j_1 1} + \varepsilon_{\alpha_2},\\
&(a_{j_1j_1\alpha_1} = a_{j_1j_1 1} + \varepsilon_{\alpha_1}, a_{j_1j_1\alpha_2} = a_{j_1j_1 1} - \varepsilon_{\alpha_2}),\\
&\varepsilon_{\alpha_1}, \varepsilon_{\alpha_2} > 0,\ \alpha_i = 1, 2, \ldots, p\,(i = 1, 2);\\
&a_{j_1j_2\beta_1} = a_{j_1j_2 1} - \varepsilon_{\beta_1}, a_{j_1j_2\beta_2} = a_{j_1j_2 1} + \varepsilon_{\beta_2},\\
&(a_{j_1j_2\beta_1} = a_{j_1j_2 1} + \varepsilon_{\beta_1}, a_{j_1j_2\beta_2} = a_{j_1j_2 1} - \varepsilon_{\beta_2}),\\
&\varepsilon_{s_1}, \varepsilon_{s_2} > 0,\ s_i = 1, 2, \ldots, q\,(i = 1, 2).
\end{aligned}$$

In summary, from cases (IV_1)–(IV_3), the 1-dimensional flow of $(x^*_{j_1}, x^*_{j_2}) = (a_{j_1j_1 1}, a_{j_1j_2 1})$ for $m = 2m_1$ and $n = 2n_1$ has the bifurcation properties as stated in (viii_{1a})–(viii_{1d}) through Eqs. (1.134–1.149).

Chapter 3
Singular Flows Bifurcations and Networks

In this chapter, the singular 1-dimensional flows in polynomial systems will be illustrated to help one understand the global bifurcations for networks of the hyperbolic and hyperbolic-secant flows. There are four classes of global bifurcations, including singular hyperbolic flows, singular hyperbolic-to-hyperbolic-secant flows, singular increasing-to-decreasing source and sink, singular increasing and decreasing-inflection saddle flows. The corresponding global bifurcations generates hyperbolic and hyperbolic-secant flow networks.

3.1 $[(2m_1+1), (2n_1+1)]$-Hyperbolic and Hyperbolic-Secant Flows

Consider a polynomial dynamical system as

$$\begin{aligned} \dot{x}_2 &= a_{220}, \\ \dot{x}_1 &= a_{110}(x_1-a_{111})^{2m_1+1}(x_2-a_{121})^{2n_1+1}. \end{aligned} \tag{3.1}$$

The first integral manifold is

$$\begin{aligned} &-\frac{1}{2m_1}a_{220}\Big[\frac{1}{(x_1-a_{111})^{2m_1}}-\frac{1}{(x_{10}-a_{111})^{2m_1}}\Big] \\ &\quad=\frac{1}{2n_1+2}a_{110}\big[(x_2-a_{121})^{2n_1+2}-(x_2-a_{121})^{2n_1+2}\big] \\ &\text{for } 2m_1+1\neq 1, \\ &a_{220}\ln\frac{|x_1-a_{111}|}{|x_{10}-a_{111}|}=\frac{1}{2n_1+2}a_{110}\big[(x_2-a_{121})^{2n_1+2}-(x_2-a_{121})^{2n_1+2}\big] \end{aligned}$$

A. C. J. Luo, *Two-dimensional Constant and Product Polynomial Systems*,
https://doi.org/10.1007/978-981-96-5515-1_3

$$\text{for } 2m_1 + 1 = 1. \tag{3.2}$$

For $(x_1^*, x_2^*) = (a_{111}, a_{121})$ with $\dot{x}_1 = 0$, the $[(2m_1 + 1), (2n_1 + 1)]$-order singularity exists with $\dot{x}_2 = a_{220} \neq 0$. For $(x_1^*, x_2^*) = (a_{111}, a_{121})$, the 1-dimensional flow direction is determined by $\dot{x}_2 \neq 0$. Thus,

$$
\begin{aligned}
&(a_{111}, a_{121}) = \underbrace{([(2m_1{+}1),\,(2n_1+1)]\text{DP:UP,pF})}_{[(2m_1+1),(2n_1+1)]\text{-hyperbolic flow }(+)},\\
&\text{for } a_{110} > 0,\, a_{220} > 0;\\
&(a_{111}, a_{121}) = \underbrace{([(2m_1{+}1),\,(2n_1+1)]\text{UP:DP,pF})}_{[(2m_1+1),(2n_1+1)]\text{-hyperbolic-secant flow }(+)},\\
&\text{for } a_{110} < 0,\, a_{220} > 0;\\
&(a_{111}, a_{121}) = \underbrace{([(2m_1{+}1),\,(2n_1+1)]\text{UP:DP,nF})}_{[(2m_1+1),(2n_1+1)]\text{-hyperbolic-secant flow }(-)},\\
&\text{for } a_{110} > 0,\, a_{220} < 0;\\
&(a_{111}, a_{121}) = \underbrace{([(2m_1{+}1),\,(2n_1+1)]\text{DP:UP,nF})}_{[(2m_1+1),(2n_1+1)]\text{-hyperbolic flow }(-)},\\
&\text{for } a_{110} < 0,\, a_{220} < 0.
\end{aligned}
\tag{3.3}
$$

The $([(2m_1{+}1), (2n_1 + 1)]$DP:UP,pF) is for a positive hyperbolic flow. The $([(2m_1{+}1), (2n_1 + 1)]$UP:DP,pF) is for a positive hyperbolic-secant flow. The $([(2m_1{+}1), (2n_1 + 1)]$UP:DP,nF) is for a negative hyperbolic-secant flow. The $([(2m_1{+}1), (2n_1 + 1)]$DP:UP,nF) is for a negative hyperbolic flow. The acronyms "UP" and "DP" are for up-parabola and down-parabola, respectively. "pF" and "nF" are for positive and negative flows, respectively. DP:UP is for a hyperbolic flow, and UP:DP is for a hyperbolic-secant flow. The singular hyperbolic and hyperbolic-secant flows are illustrated in Fig. 3.1. Through such singular flows, the corresponding global bifurcations generate from a hyperbolic (hyperbolic-secant) flow to a network of hyperbolic and hyperbolic-secant flows.

From the above bifurcations, there is a non-singular system with $(2m_1{+}1) \times (2n_1{+}1)$-hyperbolic and hyperbolic-secant flows network as

$$
\begin{aligned}
\dot{x}_1 &= a_{110} \prod_{s=1}^{2m_1+1} (x_1 - a_{11s}) \prod_{l=1}^{2n_1+1} (x_2 - a_{12l}),\\
\dot{x}_2 &= a_{220}.
\end{aligned}
\tag{3.4}
$$

The first integral manifold is given by

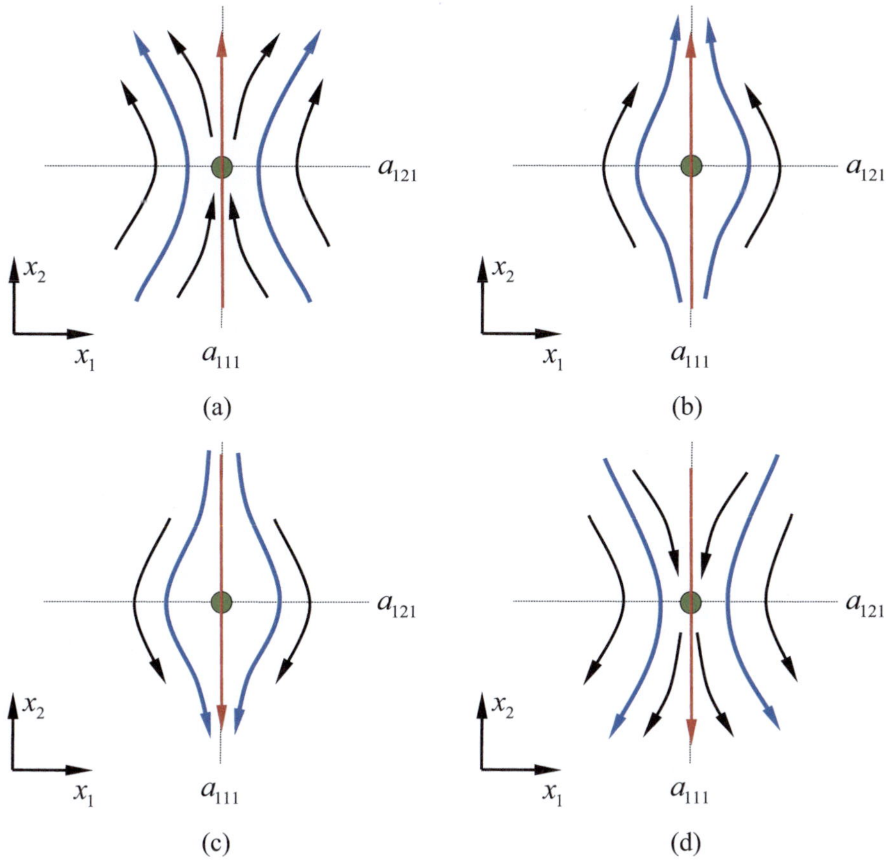

Fig. 3.1 Phase portraits of the ([2m_1 + 1], [2n_1 + 1])-hyperbolic and hyperbolic-secant singular flows of $(x_1^*, x_2^*) = (a_{111}, a_{121})$ in the x_1-direction and constant speed in the x_2-direction for 2-dimensional polynomial systems: **a** positive singular hyperbolic flow ($a_{110} > 0, a_{220} > 0$), **b** positive singular hyperbolic-secant flow ($a_{110} < 0, a_{220} > 0$), **c** negative singular hyperbolic-secant flow ($a_{110} > 0, a_{220} < 0$), **d** negative singular hyperbolic flow ($a_{110} < 0, a_{220} < 0$)

$$
\begin{aligned}
&a_{j_2j_20}\sum_{\alpha_1=1}^{2m_1+1}\frac{1}{\prod_{\alpha_2=1,\alpha_2\neq\alpha_1}^{2m_1+1}(a_{j_1j_1\alpha_1}-a_{j_1j_1\alpha_2})}\ln\frac{|x_{j_1}-a_{j_1j_1\alpha_1}|}{|x_{j_10}-a_{j_1j_1\alpha_1}|}\\
&=a_{j_1j_10}\Big\{\frac{1}{2n_1+2}\big[(x_{j_2}-a_{j_1j_2\beta_1})^{2n_1+2}-(x_{j_20}-a_{j_1j_2\beta_1})^{2n_1+2}\big]\\
&+\sum_{k=1}^{2n_1}\frac{1}{2n_1-k+2}b_{j_1j_2k}\big[(x_{j_2}-a_{j_1j_2\beta_1})^{2n_1-k+2}-(x_{j_20}-a_{j_1j_2\beta_1})^{2n_1-k+2}\big]\Big\}.
\end{aligned}
\tag{3.5}
$$

where

$$
\begin{aligned}
b_{j_1j_21} &= \sum_{\beta_2=1,\beta_2\neq\beta_1}^{2n_1+1} (a_{j_1j_2\beta_1} - a_{j_1j_2\beta_2}),\\
b_{j_1j_22} &= \sum_{\substack{\beta_2,\beta_3=1;\beta_2,\beta_3\neq\beta_1\\(\beta_2<\beta_3)}}^{2n_1+1} \prod_{r=2}^{3} (a_{j_1j_2\beta_1} - a_{j_1j_2\beta_r}), \ldots,\\
b_{j_1j_2k} &= \sum_{\substack{\beta_2,\beta_3,\ldots,\beta_{k+1}=1;\\ \beta_2,\beta_3,\ldots,\beta_{k+1}\neq\beta_1\\(\beta_2<\beta_3<\ldots<\beta_{k+1})}}^{2n_1+1} \prod_{r=2}^{k+1} (a_{j_1j_2\beta_1} - a_{j_1j_2\beta_r}), \ldots,\\
b_{j_1j_2(2n_1)} &= \prod_{\beta_1=1,\beta_2\neq s_1}^{2n_1+1} (a_{j_1j_2\beta_1} - a_{j_1j_2\beta_2}).
\end{aligned}
\tag{3.6}
$$

Phase portraits for the 2-dimensional systems with the hyperbolic and hyperbolic-secant flows at $(x^*_{j_1}, x^*_{j_2}) = (a_{j_1j_1\alpha}, a_{j_1j_2\beta})$ ($\alpha \in \{1, 2, \ldots, 2m_1 + 1\}$, $\beta \in \{1, 2, \ldots, 2n_1 + 1\}$) for the x_{j_1}-direction are presented in Fig. 3.2a–d for ($a_{110} > 0, a_{220} > 0$), ($a_{110} < 0, a_{220} > 0$), ($a_{110} > 0, a_{220} < 0$) and ($a_{110} < 0, a_{220} < 0$). The equilibriums of $(x^*_{j_1}, x^*_{j_2}) = (a_{j_1j_1\alpha}, a_{j_1j_2\beta})$ ($\alpha \in \{1, 2, \ldots, 2m_1 + 1\}$, $\beta \in \{1, 2, \ldots, 2n_1 + 1\}$) for the x_{j_1}-direction are for hyperbolic and hyperbolic-secant flow network matrices, which are generated from hyperbolic or hyperbolic-secant flow to the corresponding network.

- *For* $a_{110} > 0$ and $a_{220} > 0$,

$$
\bigcup_{\alpha=1}^{2m_1+1}\bigcup_{\beta=1}^{2n_1+1} \underbrace{(a_{11\alpha}, a_{12\beta})}_{\text{(XX, pF)}}
= \begin{bmatrix}
\underbrace{\text{(DP : UP, pF)}}_{\text{hyperbolic flow }(+)} & \cdots & \underbrace{\text{(UP : DP, pF)}}_{\text{hyperbolic-secant flow }(+)} & \underbrace{\text{(DP : UP, pF)}}_{\text{hyperbolic flow }(+)}\\
\underbrace{\text{(UP : DP, pF)}}_{\text{hyperbolic-secant flow }(+)} & \cdots & \underbrace{\text{(DP : UP, pF)}}_{\text{hyperbolic flow }(+)} & \underbrace{\text{(UP : DP, pF)}}_{\text{hyperbolic-secant flow }(+)}\\
\vdots & \vdots & \vdots & \vdots\\
\underbrace{\text{(DP : UP, pF)}}_{\text{hyperbolic flow }(+)} & \cdots & \underbrace{\text{(UP : DP, pF)}}_{\text{hyperbolic-secant flow }(+)} & \underbrace{\text{(DP : UP, pF)}}_{\text{hyperbolic flow }(+)}
\end{bmatrix}_{(2n_1+1)\times(2m_1+1)}
\tag{3.7}
$$

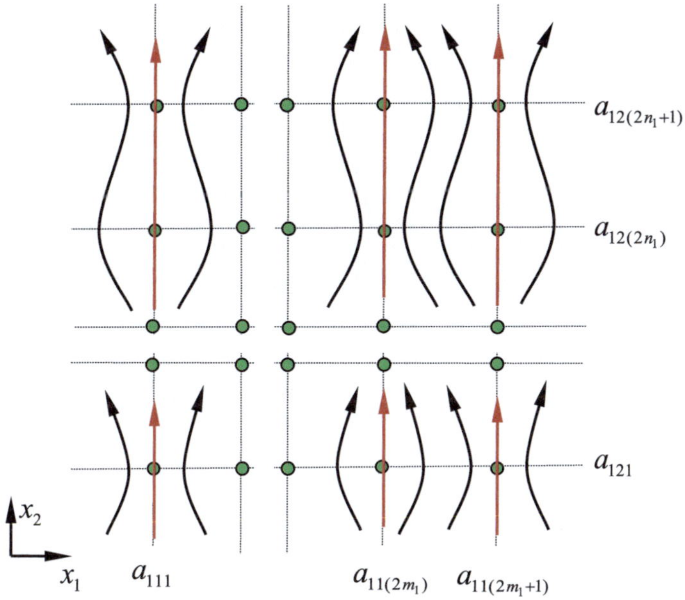

(a)

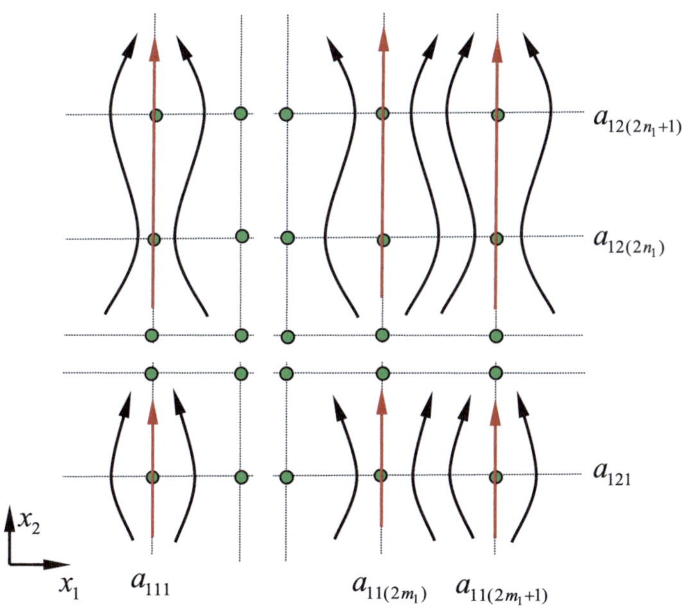

(b)

Fig. 3.2 Phase portraits for 2-dimensional systems with $x_1^* = a_{111}, a_{112}, \ldots, a_{11(2m_1+1)}$ and $x_2^* = a_{121}, a_{122}, \ldots, a_{12(2n_1+1)}$. The four networks of hyperbolic and hyperbolic-secant flow: **a** ($a_{110} > 0, a_{220} > 0$), **b** ($a_{110} < 0, a_{220} > 0$), **c** ($a_{110} > 0, a_{220} < 0$), **d** ($a_{110} < 0, a_{220} < 0$)

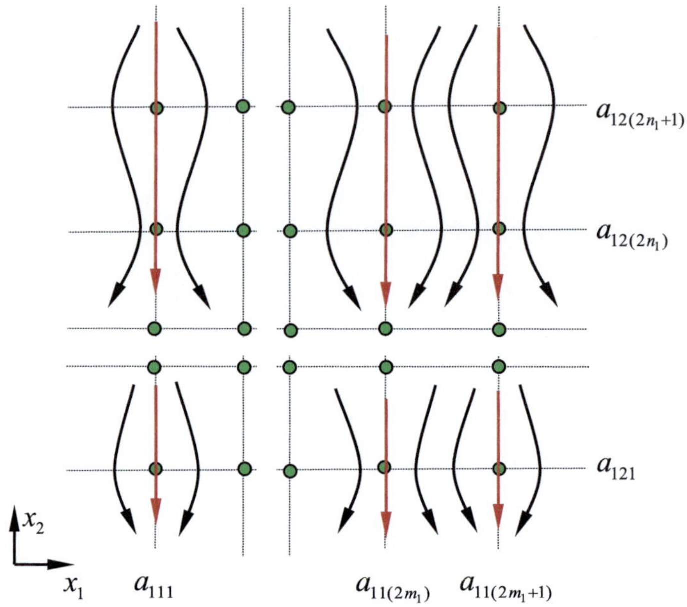

(c)

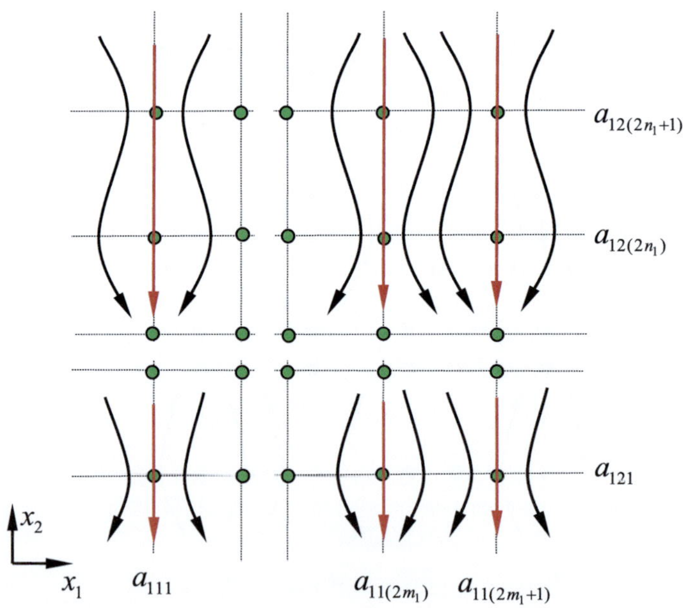

(d)

Fig. 3.2 (continued)

- *For* $a_{110} < 0$ and $a_{220} > 0$,

$$\bigcup_{\alpha=1}^{2m_1+1}\bigcup_{\beta=1}^{2n_1+1}\underbrace{(a_{11\alpha}, a_{12\beta})}_{\text{(XX, pF)}}$$
$$=\begin{bmatrix}\underbrace{\text{(UP : DP, pF)}}_{\text{hyperbolic-secant flow }(+)} & \cdots & \underbrace{\text{(DP : UP, pF)}}_{\text{hyperbolic flow }(+)} & \underbrace{\text{(UP : DP, pF)}}_{\text{hyperbolic-secant flow }(+)}\\ \underbrace{\text{(DP : UP, pF)}}_{\text{hyperbolic flow }(+)} & \cdots & \underbrace{\text{(UP : DP, pF)}}_{\text{hyperbolic-secant flow }(+)} & \underbrace{\text{(DP : UP, pF)}}_{\text{hyperbolic flow }(+)}\\ \vdots & \vdots & \vdots & \vdots\\ \underbrace{\text{(UP : DP, pF)}}_{\text{hyperbolic-secant flow }(+)} & \cdots & \underbrace{\text{(DP : UP, pF)}}_{\text{hyperbolic flow }(+)} & \underbrace{\text{(UP : DP, pF)}}_{\text{hyperbolic-secant flow }(+)}\end{bmatrix}_{(2n_1+1)\times(2m_1+1)} \tag{3.8}$$

- *For* $a_{110} > 0$ and $a_{220} < 0$,

$$\bigcup_{\alpha=1}^{2m_1+1}\bigcup_{\beta=1}^{2n_1+1}\underbrace{(a_{11\alpha}, a_{12\beta})}_{\text{(XX, nF)}}$$
$$=\begin{bmatrix}\underbrace{\text{(UP : DP, nF)}}_{\text{hyperbolic-secant flow }(-)} & \cdots & \underbrace{\text{(DP : UP, nF)}}_{\text{hyperbolic flow }(-)} & \underbrace{\text{(UP : DP, nF)}}_{\text{hyperbolic-secant flow }(-)}\\ \underbrace{\text{(DP : UP, nF)}}_{\text{hyperbolic flow }(-)} & \cdots & \underbrace{\text{(UP : DP, nF)}}_{\text{hyperbolic-secant flow }(-)} & \underbrace{\text{(DP : UP, nF)}}_{\text{hyperbolic flow }(-)}\\ \vdots & \vdots & \vdots & \vdots\\ \underbrace{\text{(UP : DP, nF)}}_{\text{hyperbolic-secant flow }(-)} & \cdots & \underbrace{\text{(DP : UP, nF)}}_{\text{hyperbolic flow }(-)} & \underbrace{\text{(UP : DP, nF)}}_{\text{hyperbolic-secant flow }(-)}\end{bmatrix}_{(2n_1+1)\times(2m_1+1)} \tag{3.9}$$

- *For* $a_{110} < 0$ and $a_{220} < 0$,

$$\bigcup_{\alpha=1}^{2m_1+1}\bigcup_{\beta=1}^{2n_1+1}\underbrace{(a_{11\alpha}, a_{12\beta})}_{\text{(XX, nF)}}$$
$$=\begin{bmatrix}\underbrace{\text{(DP : UP, nF)}}_{\text{hyperbolic flow }(-)} & \cdots & \underbrace{\text{(UP : DP, nF)}}_{\text{hyperbolic-secant flow }(-)} & \underbrace{\text{(DP : UP, nF)}}_{\text{hyperbolic flow }(-)}\\ \underbrace{\text{(UP : DP, nF)}}_{\text{hyperbolic-secant flow }(-)} & \cdots & \underbrace{\text{(DP : UP, nF)}}_{\text{hyperbolic flow }(-)} & \underbrace{\text{(UP : DP, nF)}}_{\text{hyperbolic-secant flow }(-)}\\ \vdots & \vdots & \vdots & \vdots\\ \underbrace{\text{(DP : UP, nF)}}_{\text{hyperbolic flow }(-)} & \cdots & \underbrace{\text{(UP : DP, nF)}}_{\text{hyperbolic-secant flow }(-)} & \underbrace{\text{(DP : UP, nF)}}_{\text{hyperbolic flow }(-)}\end{bmatrix}_{(2n_1+1)\times(2m_1+1)} \tag{3.10}$$

The lines of $x_1^* = a_{11\alpha}$ $(\alpha \in \{1, 2, \ldots, 2m_1 + 1\})$ *are directrix flows.*

3.2 $[(2m_1), (2n_1 + 1)]$-Hyperbolic-to-Hyperbolic-Secant Flows

Consider a polynomial dynamical system as

$$\begin{aligned} \dot{x}_2 &= a_{220}, \\ \dot{x}_1 &= a_{110}(x_1 - a_{111})^{2m_1}(x_2 - a_{121})^{2n_1+1}. \end{aligned} \tag{3.11}$$

The first integral manifold is

$$\begin{aligned} &-\frac{1}{2m_1 - 1} a_{220}\Big[\frac{1}{(x_1 - a_{111})^{2m_1-1}} - \frac{1}{(x_{10} - a_{111})^{2m_1-1}}\Big] \\ &= \frac{1}{2n_1 + 2} a_{110}\big[(x_2 - a_{121})^{2n_1+2} - (x_2 - a_{121})^{2n_1+2}\big] \\ &\text{for } 2m_1 - 1 \neq 0. \end{aligned} \tag{3.12}$$

For $(x_1^*, x_2^*) = (a_{111}, a_{121})$ with $\dot{x}_1 = 0$, the $[(2m_1), (2n_1+1)]$-order singularity exists with $\dot{x}_2 = a_{220} \neq 0$. For $(x_1^*, x_2^*) = (a_{111}, a_{121})$, the 1-dimensional flow direction is determined by $\dot{x}_2 \neq 0$. Thus,

$$\begin{aligned} &(a_{111}, a_{121}) = \underbrace{([(2m_1), (2n_1+1)]\text{UP:UP,pF})}_{[(2m_1),(2n_1+1)]\text{-hyperbolic-secant-to-hyperbolic flow }(+)}, \\ &\text{for } a_{110} > 0, a_{220} > 0; \\ &(a_{111}, a_{121}) = \underbrace{([(2m_1), (2n_1+1)]\text{UP:DP,pF})}_{[(2m_1),(2n_1+1)]\text{-hyperblic-to-hyperbolic-secant flow }(+)}, \\ &\text{for } a_{110} < 0, a_{220} > 0; \\ &(a_{111}, a_{121}) = \underbrace{([(2m_1), (2n_1+1)]\text{UP:DP,nF})}_{[(2m_1),(2n_1+1)]\text{-hyperbolic-to-hyperbolic-secant flow }(-)}, \\ &\text{for } a_{110} > 0, a_{220} < 0; \\ &(a_{111}, a_{121}) = \underbrace{([(2m_1), (2n_1+1)]\text{DP:UP,nF})}_{[(2m_1),(2n_1+1)]\text{-hyperbolic-secant-to-hyperbolic flow }(-)}, \\ &\text{for } a_{110} < 0, a_{220} < 0. \end{aligned} \tag{3.13}$$

The $([(2m_1), (2n_1 + 1)]$UP:UP,pF) is for a positive hyperbolic-secant-to-hyperbolic flow. The $([(2m_1), (2n_1 + 1)]$DP:DP,pF) is for a positive hyperbolic-to-hyperbolic-secant flow. The $([(2m_1), (2n_1 + 1)]$UP:UP,nF) is for a negative

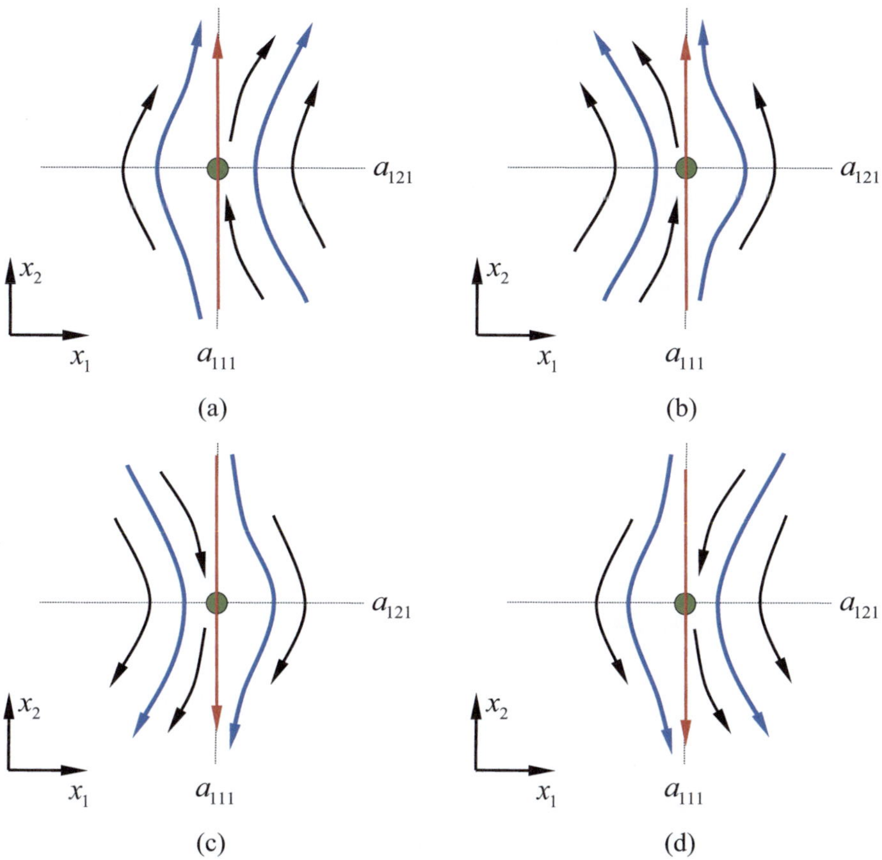

Fig. 3.3 Phase portraits of the ([$2m_1$], [$2n_1+1$])-hyperbolic-secant-to-hyperbolic and hyperbolic-to-hyperbolic-secant singular flows of $(x_1^*, x_2^*) = (a_{111}, a_{121})$ in the x_1-direction and constant speed in the x_2-direction for 2-dimensional polynomial systems: **a** positive singular hyperbolic flow ($a_{110} > 0, a_{220} > 0$), **b** positive singular hyperbolic-secant flow ($a_{110} < 0, a_{220} > 0$), **c** negative singular hyperbolic-secant flow ($a_{110} > 0, a_{220} < 0$), **d** negative singular hyperbolic flow ($a_{110} < 0, a_{220} < 0$)

hyperbolic-to-hyperbolic-secant flow. The ([($2m_1$), ($2n_1$ + 1)]UP:UP,nF) is for a negative hyperbolic-to-hyperbolic flow. UP:UP is for a hyperbolic-secant-to-hyperbolic flow, and DP:DP is for a hyper-secant-to-hyperbolic-secant flow. The singular hyperbolic and hyperbolic-secant flows are illustrated in Fig. 3.3. Through such singular flows, the corresponding global bifurcations generate from a hyperbolic (hyperbolic-secant) flow to a network of hyperbolic and hyperbolic-secant flows.

From the above bifurcations, there is a non-singular system with $(2m_1) \times (2n_1 + 1)$-hyperbolic and hyperbolic-secant flows as

$$\begin{aligned}
\dot{x}_1 &= a_{110}\prod_{s=1}^{2m_1}(x_1 - a_{11s})\prod_{l=1}^{2n_1+1}(x_2 - a_{12l}),\\
\dot{x}_2 &= a_{220}.
\end{aligned} \tag{3.14}$$

The first integral manifold is given by

$$\begin{aligned}
&a_{j_2j_20}\sum_{\alpha_1=1}^{2m_1}\frac{1}{\prod_{\alpha_2=1,\alpha_2\neq\alpha_1}^{2m_1}(a_{j_1j_1\alpha_1} - a_{j_1j_1\alpha_2})}\ln\frac{|x_{j_1} - a_{j_1j_1\alpha_1}|}{|x_{j_10} - a_{j_1j_1\alpha_1}|}\\
&= a_{j_1j_10}\Big\{\frac{1}{2n_1+2}\big[(x_{j_2} - a_{j_1j_2\beta_1})^{2n_1+2} - (x_{j_20} - a_{j_1j_2\beta_1})^{2n_1+2}\big]\\
&+\sum_{k=1}^{2n_1}\frac{1}{2n_1-k+1}b_{j_1j_2k}\big[(x_{j_2} - a_{j_1j_2\beta_1})^{2n_1-k+2} - (x_{j_20} - a_{j_1j_2\beta_1})^{2n_1-k+2}\big]\Big\}.
\end{aligned} \tag{3.15}$$

where

$$\begin{aligned}
b_{j_1j_21} &= \sum_{\beta_2=1,\beta_2\neq\beta_1}^{2n_1+1}(a_{j_1j_2\beta_1} - a_{j_1j_2\beta_2}),\\
b_{j_1j_22} &= \sum_{\substack{\beta_2,\beta_3=1;\beta_2,\beta_3\neq\beta_1\\(\beta_2<\beta_3)}}^{2n_1+1}\prod_{r=2}^{3}(a_{j_1j_2\beta_1} - a_{j_1j_2\beta_r}),\ldots,\\
b_{j_1j_2k} &= \sum_{\substack{\beta_2,\beta_3,\ldots,\beta_{k+1}=1;\\ \beta_2,\beta_3,\ldots,\beta_{k+1}\neq\beta_1\\(\beta_2<\beta_3<\ldots<\beta_{k+1})}}^{2n_1+1}\prod_{r=2}^{k+1}(a_{j_1j_2\beta_1} - a_{j_1j_2\beta_r}),\ldots,\\
b_{j_1j_2(2n_1)} &= \prod_{\beta_1=1,\beta_2\neq s_1}^{2n_1+1}(a_{j_1j_2\beta_1} - a_{j_1j_2\beta_2}).
\end{aligned} \tag{3.16}$$

Phase portraits for the 2-dimensional systems with the hyperbolic and hyperbolic-secant flows at $(x_{j_1}^*, x_{j_2}^*) = (a_{j_1j_1\alpha}, a_{j_1j_2\beta})$ ($\alpha \in \{1, 2, \ldots, 2m_1\}$, $\beta \in \{1, 2, \ldots, 2n_1 + 1\}$) for the x_{j_1}-direction are presented in Fig. 3.4a–d for ($a_{110} > 0, a_{220} > 0$), ($a_{110} < 0, a_{220} > 0$), ($a_{110} > 0, a_{220} < 0$) and ($a_{110} < 0, a_{220} < 0$). The equilibriums of $(x_{j_1}^*, x_{j_2}^*) = (a_{j_1j_1\alpha}, a_{j_1j_2\beta})$ ($\alpha \in \{1, 2, \ldots, 2m_1\}, \beta \in \{1, 2, \ldots, 2n_1 + 1\}$) for the x_{j_1}-direction are for hyperbolic and hyperbolic-secant flow network matrices, which are generated from a parabola flow to the corresponding network.

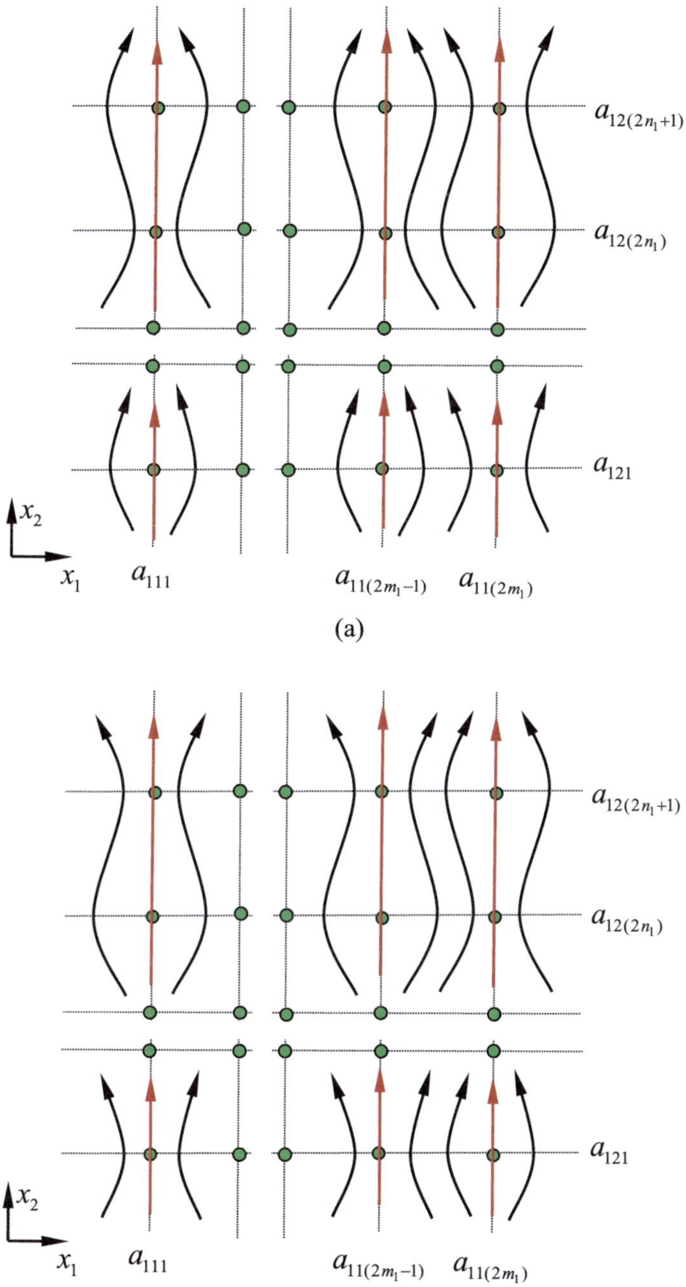

Fig. 3.4 Phase portraits for 2-dimensional systems with $x_1^* = a_{111}, a_{112}, \ldots, a_{11(2m_1)}$ and $x_2^* = a_{121}, a_{122}, \ldots, a_{12(2n_1+1)}$. The four networks of hyperbolic and hyperbolic-secant flow: **a** ($a_{110} > 0, a_{220} > 0$), **b** ($a_{110} < 0, a_{220} > 0$), **c** ($a_{110} > 0, a_{220} < 0$), **d** ($a_{110} < 0, a_{220} < 0$)

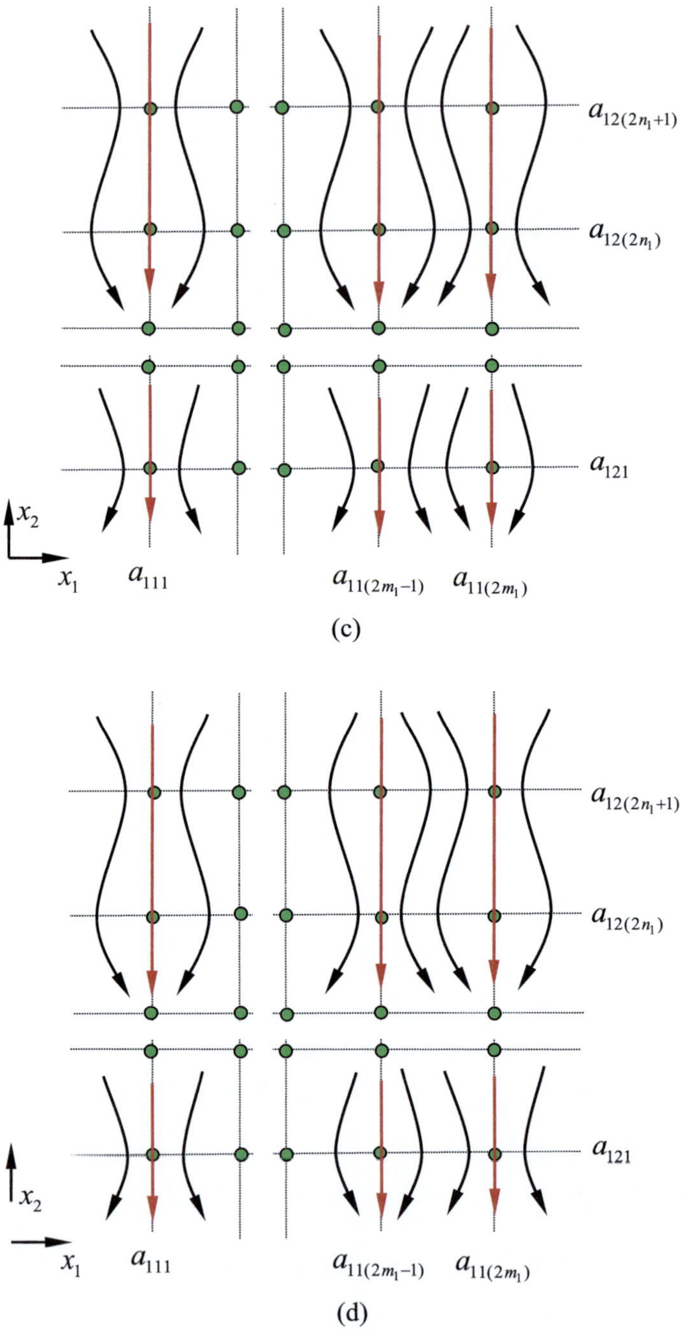

Fig. 3.4 (continued)

- *For* $a_{110} > 0$ *and* $a_{220} > 0$,

$$\bigcup_{\alpha=1}^{2m_1}\bigcup_{\beta=1}^{2n_1+1}\underbrace{(a_{11\alpha},a_{12\beta})}_{\text{(XX, pF)}} = \begin{bmatrix} \underbrace{\text{(UP : DP, pF)}}_{\text{hyperbolic-secant flow (+)}} & \cdots & \underbrace{\text{(UP : DP, pF)}}_{\text{hyperbolic-secant flow (+)}} & \underbrace{\text{(DP : UP, pF)}}_{\text{hyperbolic flow (+)}} \\ \underbrace{\text{(DP : UP, pF)}}_{\text{hyperbolic flow (+)}} & \cdots & \underbrace{\text{(DP : UP, pF)}}_{\text{hyperbolic flow (+)}} & \underbrace{\text{(UP : DP, pF)}}_{\text{hyperbolic-secant flow (+)}} \\ \vdots & \vdots & \vdots & \vdots \\ \underbrace{\text{(UP : DP, pF)}}_{\text{hyperbolic-secant flow (+)}} & \cdots & \underbrace{\text{(UP : DP, pF)}}_{\text{hyperbolic-secant flow (+)}} & \underbrace{\text{(DP : UP, pF)}}_{\text{hyperbolic flow (+)}} \end{bmatrix}_{(2n_1+1)\times(2m_1)} \tag{3.17}$$

- *For* $a_{110} < 0$ and $a_{220} > 0$,

$$\bigcup_{\alpha=1}^{2m_1}\bigcup_{\beta=1}^{2n_1+1}\underbrace{(a_{11\alpha},a_{12\beta})}_{\text{(XX, pF)}} = \begin{bmatrix} \underbrace{\text{(DP : UP, pF)}}_{\text{hyperbolic flow (+)}} & \cdots & \underbrace{\text{(DP : UP, pF)}}_{\text{hyperbolic flow (+)}} & \underbrace{\text{(UP : DP, pF)}}_{\text{hyperbolic-secant flow (+)}} \\ \underbrace{\text{(UP : DP, pF)}}_{\text{hyperbolic-secant flow (+)}} & \cdots & \underbrace{\text{(UP : DP, pF)}}_{\text{hyperbolic-secant flow (+)}} & \underbrace{\text{(DP : UP, pF)}}_{\text{hyperbolic flow (+)}} \\ \vdots & \vdots & \vdots & \vdots \\ \underbrace{\text{(DP : UP, pF)}}_{\text{hyperbolic flow (+)}} & \cdots & \underbrace{\text{(DP : UP, pF)}}_{\text{hyperbolic flow (+)}} & \underbrace{\text{(UP : DP, pF)}}_{\text{hyperbolic-secant flow (+)}} \end{bmatrix}_{(2n_1+1)\times(2m_1)} \tag{3.18}$$

- *For* $a_{110} > 0$ and $a_{220} < 0$,

$$\bigcup_{\alpha=1}^{2m_1}\bigcup_{\beta=1}^{2n_1+1}\underbrace{(a_{11\alpha},a_{12\beta})}_{\text{(XX, nF)}} = \begin{bmatrix} \underbrace{\text{(DP : UP, nF)}}_{\text{hyperbolic flow (−)}} & \cdots & \underbrace{\text{(DP : UP, nF)}}_{\text{hyperbolic flow (−)}} & \underbrace{\text{(UP : DP, nF)}}_{\text{hyperbolic-secant flow (−)}} \\ \underbrace{\text{(UP : DP, nF)}}_{\text{hyperbolic-secant flow (−)}} & \cdots & \underbrace{\text{(UP : DP, nF)}}_{\text{hyperbolic-secant flow (−)}} & \underbrace{\text{(DP : UP, nF)}}_{\text{hyperbolic flow (−)}} \\ \vdots & \vdots & \vdots & \vdots \\ \underbrace{\text{(DP : UP, nF)}}_{\text{hyperbolic flow (−)}} & \cdots & \underbrace{\text{(DP : UP, nF)}}_{\text{hyperbolic flow (−)}} & \underbrace{\text{(UP : DP, nF)}}_{\text{hyperbolic-secant flow (−)}} \end{bmatrix}_{(2n_1+1)\times(2m_1+1)} \tag{3.19}$$

- *For* $a_{110} < 0$ and $a_{220} < 0$,

$$\bigcup_{\alpha=1}^{2m_1}\bigcup_{\beta=1}^{2n_1+1}\underbrace{(a_{11\alpha},a_{12\beta})}_{\text{(XX, nF)}}$$

$$=\begin{bmatrix}\underbrace{\text{(UP : DP, nF)}}_{\text{hyperbolic-secant flow }(-)} & \cdots & \underbrace{\text{(UP : DP, nF)}}_{\text{hyperbolic-secant flow }(-)} & \underbrace{\text{(DP : UP, nF)}}_{\text{hyperbolic flow }(-)}\\ \underbrace{\text{(DP : UP, nF)}}_{\text{hyperbolic flow }(-)} & \cdots & \underbrace{\text{(DP : UP, nF)}}_{\text{hyperbolic flow }(-)} & \underbrace{\text{(UP : DP, nF)}}_{\text{hyperbolic-secant flow }(-)}\\ \vdots & \vdots & \vdots & \vdots\\ \underbrace{\text{(UP : DP, nF)}}_{\text{hyperbolic-secant flow }(-)} & \cdots & \underbrace{\text{(UP : DP, nF)}}_{\text{hyperbolic-secant flow }(-)} & \underbrace{\text{(DP : UP, nF)}}_{\text{hyperbolic flow }(-)}\end{bmatrix}_{(2n_1+1)\times(2m_1)} \tag{3.20}$$

The lines of $x_1^* = a_{11\alpha}$ $(\alpha \in \{1, 2, \ldots, 2m_1\})$ *are directrix flows.*

3.3 $[(2m_1 + 1), (2n_1)]$-Inflection Source and Sink Flows

Consider a polynomial dynamical system as

$$\begin{aligned}\dot{x}_2 &= a_{220},\\ \dot{x}_1 &= a_{110}(x_1 - a_{111})^{2m_1+1}(x_2 - a_{121})^{2n_1}.\end{aligned} \tag{3.21}$$

The first integral manifold is

$$\begin{aligned}&-\frac{1}{2m_1}a_{220}\Big[\frac{1}{(x_1 - a_{111})^{2m_1}} - \frac{1}{(x_{10} - a_{111})^{2m_1}}\Big]\\ &\quad= \frac{1}{2n_1 + 1}a_{110}\big[(x_2 - a_{121})^{2n_1+1} - (x_2 - a_{121})^{2n_1+1}\big]\\ &\text{for } 2m_1 + 1 \neq 1,\\ &a_{220}\ln\frac{|x_1 - a_{111}|}{|x_{10} - a_{111}|} = \frac{1}{2n_1 + 1}a_{110}\big[(x_2 - a_{121})^{2n_1+1} - (x_2 - a_{121})^{2n_1+1}\big]\\ &\text{for } 2m_1 + 1 = 1.\end{aligned} \tag{3.22}$$

For $(x_1^*, x_2^*) = (a_{111}, a_{121})$ with $\dot{x}_1 = 0$, the $[(2m_1+1), (2n_1)]$-order singularity exists with $\dot{x}_2 = a_{220} \neq 0$. For $(x_1^*, x_2^*) = (a_{111}, a_{121})$, the 1-dimensional flow direction is determined by $\dot{x}_2 \neq 0$. Thus,

$$
\begin{aligned}
&(a_{111}, a_{121}) = \underbrace{([(2m_1+1),\ (2n_1)]_{\mathrm{DI:II}}\mathrm{SO,pF})}_{[(2m_1+1),(2n_1)]\text{-decreasing-to-increasing inflection source flow }(+)},\\
&\text{for } a_{110} > 0, a_{220} > 0;\\
&(a_{111}, a_{121}) = \underbrace{([(2m_1+1),\ (2n_1)]_{\mathrm{II:DI}}\mathrm{SI,pF})}_{[(2m_1+1),(2n_1)]\text{-increasing-to-decreasing inflection sink flow }(+)},\\
&\text{for } a_{110} < 0, a_{220} > 0;\\
&(a_{111}, a_{121}) = \underbrace{([(2m_1+1),\ (2n_1)]_{\mathrm{II:DI}}\mathrm{SO,nF})}_{[(2m_1+1),(2n_1)]\text{-increasing-to-decreasing inflection source flow }(-)},\\
&\text{for } a_{110} > 0, a_{220} < 0;\\
&(a_{111}, a_{121}) = \underbrace{([(2m_1+1),\ (2n_1)]_{\mathrm{DI:II}}\mathrm{SI,pF})}_{[(2m_1+1),(2n_1)]\text{-decreasing-to-increasing inflection source flow }(-)},\\
&\text{for } a_{110} < 0, a_{220} < 0.
\end{aligned}
\tag{3.23}
$$

The $([(2m_1{+}1),\ (2n_1)]_{\mathrm{DI:II}}\mathrm{SO,pF})$ is for a positive decreasing-to-increasing-inflection source flow. The $([(2m_1{+}1),\ (2n_1)]_{\mathrm{II:DI}}\mathrm{SI,pF})$ is for a positive increasing-to-decreasing-inflection sink flow. The $([(2m_1{+}1),\ (2n_1)]_{\mathrm{II:DI}}\mathrm{SO,nF})$ is for a negative increasing-to-decreasing-inflection source flow. The $([(2m_1{+}1),\ (2n_1)]_{\mathrm{DI:II}}\mathrm{SI,\ nF})$ is for a negative decreasing-to-increasing-inflection source flow. The acronyms "II" and "DI" are for increasing and decreasing-inflection flows, respectively. "pF" and "nF" are for positive and negative flows, respectively. SO is for a source flow, and SI is for a sink flow. The singular hyperbolic and hyperbolic-secant flows are illustrated in Fig. 3.5. Through such singular flows, the corresponding global bifurcations generate from a source (sink) flow to a network of hyperbolic and hyperbolic-secant flows.

From the above bifurcations, there is a non-singular system with $(2m_1 + 1) \times (2n_1 + 1)$-hyperbolic and hyperbolic-secant flows as

$$
\begin{aligned}
&\dot{x}_1 = a_{110}\prod_{s=1}^{2m_1+1}(x_1 - a_{11s})\prod_{l=1}^{2n_1}(x_2 - a_{12l}),\\
&\dot{x}_2 = a_{220}.
\end{aligned}
\tag{3.24}
$$

The first integral manifold is given by

$$
\begin{aligned}
&a_{j_2j_20}\sum_{\alpha_1=1}^{2m_1+1}\frac{1}{\prod_{\alpha_2=1,\alpha_2\neq\alpha_1}^{2m_1+1}(a_{j_1j_1\alpha_1} - a_{j_1j_1\alpha_2})}\ln\frac{|x_{j_1} - a_{j_1j_1\alpha_1}|}{|x_{j_10} - a_{j_1j_1\alpha_1}|}\\
&= a_{j_1j_10}\Big\{\frac{1}{2n_1+1}\big[(x_{j_2} - a_{j_1j_2\beta_1})^{2n_1+1} - (x_{j_20} - a_{j_1j_2\beta_1})^{2n_1+1}\big]\\
&+\sum_{k=1}^{2n_1-1}\frac{1}{2n_1-k+1}b_{j_1j_2k}\big[(x_{j_2} - a_{j_1j_2\beta_1})^{2n_1-k+1} - (x_{j_20} - a_{j_1j_2\beta_1})^{2n_1-k+1}\big]\Big\}.
\end{aligned}
\tag{3.25}
$$

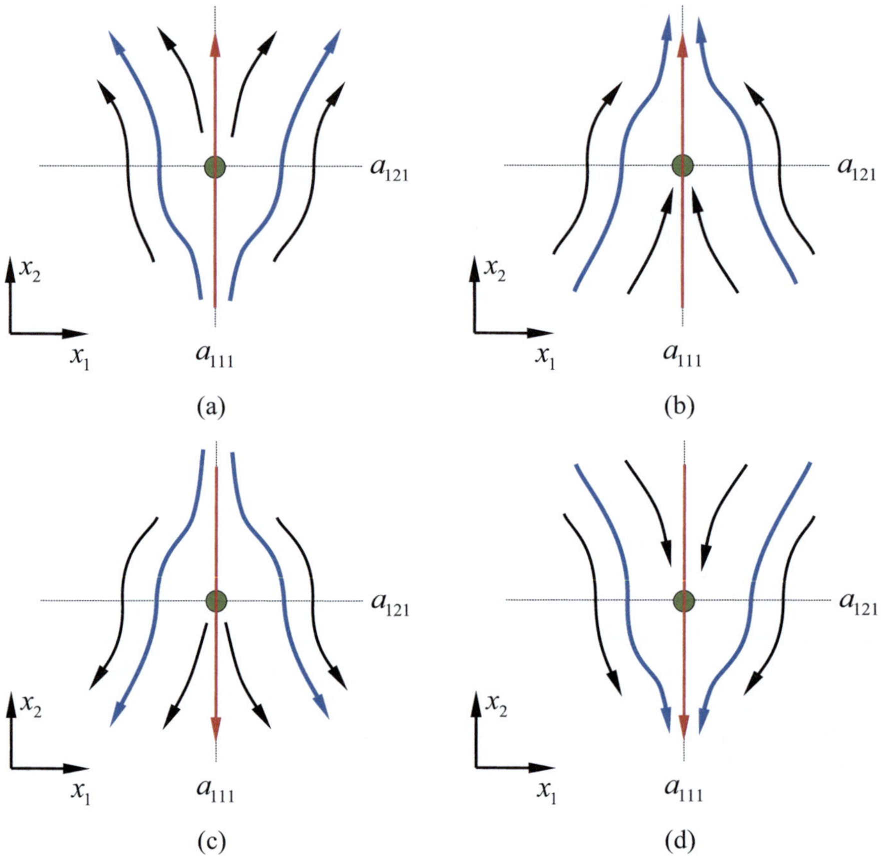

Fig. 3.5 Phase portraits of the $([2m_1 + 1], [2n_1])$-increasing-to-decreasing inflection source and sink singular flows of $(x_1^*, x_2^*) = (a_{111}, a_{121})$ in the x_1-direction and constant speed in the x_2-direction for 2-dimensional polynomial systems: **a** positive singular decreasing -to-increasing-inflection source flow ($a_{110} > 0, a_{220} > 0$), **b** positive singular increasing-to-decreasing-inflection sink flow ($a_{110} < 0, a_{220} > 0$), **c** negative singular increasing-to-decreasing-inflection source flow ($a_{110} > 0, a_{220} < 0$), **d** negative singular decreasing -to-increasing-inflection sink flow ($a_{110} < 0, a_{220} < 0$)

where

$$
\begin{aligned}
b_{j_1j_2 1} &= \sum_{\beta_2=1,\beta_2\neq\beta_1}^{2n_1} (a_{j_1j_2\beta_1} - a_{j_1j_2\beta_2}),\\
b_{j_1j_2 2} &= \sum_{\substack{\beta_2,\beta_3=1;\beta_2,\beta_3\neq\beta_1\\(\beta_2<\beta_3)}}^{2n_1} \prod_{r=2}^{3} (a_{j_1j_2\beta_1} - a_{j_1j_2\beta_r}),\ldots,\\
b_{j_1j_2 k} &= \sum_{\substack{\beta_2,\beta_3,\ldots,\beta_{k+1}=1;\\ \beta_2,\beta_3,\ldots,\beta_{k+1}\neq\beta_1\\(\beta_2<\beta_3<\cdots<\beta_{k+1})}}^{2n_1} \prod_{r=2}^{k+1} (a_{j_1j_2\beta_1} - a_{j_1j_2\beta_r}),\ldots,\\
b_{j_1j_2(2n_1-1)} &= \prod_{\beta_1=1,\beta_2\neq s_1}^{2n_1} (a_{j_1j_2\beta_1} - a_{j_1j_2\beta_2}).
\end{aligned}
\tag{3.26}
$$

Phase portraits for the 2-dimensional systems with the hyperbolic and hyperbolic-secant flows at $(x_{j_1}^*, x_{j_2}^*) = (a_{j_1j_1\alpha}, a_{j_1j_2\beta})(\alpha \in \{1, 2, \ldots, 2m_1+1\}, \beta \in \{1, 2, \ldots, 2n_1\})$ for the x_{j_1}-direction are presented in Fig. 3.6a–d for $(a_{110} > 0, a_{220} > 0)$, $(a_{110} < 0, a_{220} > 0)$, $(a_{110} > 0, a_{220} < 0)$ and $(a_{110} < 0, a_{220} < 0)$. The equilibriums of $(x_{j_1}^*, x_{j_2}^*) = (a_{j_1j_1\alpha}, a_{j_1j_2\beta})$ for the x_{j_1}-direction are for hyperbolic and hyperbolic-secant flow network matrices, which are generated from a source or sink flow to the corresponding network.

- *For* $a_{110} > 0$ and $a_{220} > 0$,

$$
\bigcup_{\alpha=1}^{2m_1+1}\bigcup_{\beta=1}^{2n_1}\underbrace{(a_{11\alpha}, a_{12\beta})}_{(\mathrm{XX,pF})}
= \begin{bmatrix}
\underbrace{(\mathrm{DP:UP,pF})}_{\text{hyperbolic flow }(+)} & \cdots & \underbrace{(\mathrm{UP:DP,pF})}_{\text{hyperbolic-secant flow }(+)} & \underbrace{(\mathrm{DP:UP,pF})}_{\text{hyperbolic flow }(+)}\\
\underbrace{(\mathrm{UP:DP,pF})}_{\text{hyperbolic-secant flow }(+)} & \cdots & \underbrace{(\mathrm{DP:UP,pF})}_{\text{hyperbolic flow }(+)} & \underbrace{(\mathrm{UP:DP,pF})}_{\text{hyperbolic-secant flow }(+)}\\
\vdots & \vdots & \vdots & \vdots\\
\underbrace{(\mathrm{UP:DP,pF})}_{\text{hyperbolic-secant flow }(+)} & \cdots & \underbrace{(\mathrm{DP:UP,pF})}_{\text{hyperbolic flow }(+)} & \underbrace{(\mathrm{UP:DP,pF})}_{\text{hyperbolic-secant flow }(+)}
\end{bmatrix}_{(2n_1)\times(2m_1+1)}
\tag{3.27}
$$

- *For* $a_{110} < 0$ and $a_{220} > 0$,

$$\bigcup_{\alpha=1}^{2m_1+1}\bigcup_{\beta=1}^{2n_1}\underbrace{(a_{11\alpha},a_{12\beta})}_{(\text{XX, pF})} = \begin{bmatrix} \underbrace{(\text{UP : DP, pF})}_{\text{hyperbolic-secant flow }(+)} & \cdots & \underbrace{(\text{DP : UP, pF})}_{\text{hyperbolic flow }(+)} & \underbrace{(\text{UP : DP, pF})}_{\text{hyperbolic-secant flow }(+)} \\ \underbrace{(\text{DP : UP, pF})}_{\text{hyperbolic flow }(+)} & \cdots & \underbrace{(\text{UP : DP, pF})}_{\text{hyperbolic-secant flow }(+)} & \underbrace{(\text{DP : UP, pF})}_{\text{hyperbolic flow }(+)} \\ \vdots & \vdots & \vdots & \vdots \\ \underbrace{(\text{DP : UP, pF})}_{\text{hyperbolic flow }(+)} & \cdots & \underbrace{(\text{UP : DP, pF})}_{\text{hyperbolic-secant flow }(+)} & \underbrace{(\text{DP : UP, pF})}_{\text{hyperbolic flow }(+)} \end{bmatrix}_{(2n_1)\times(2m_1+1)} \tag{3.28}$$

- *For* $a_{110} > 0$ and $a_{220} < 0$,

$$\bigcup_{\alpha=1}^{2m_1+1}\bigcup_{\beta=1}^{2n_1}\underbrace{(a_{11\alpha},a_{12\beta})}_{(\text{XX, nF})} = \begin{bmatrix} \underbrace{(\text{UP : DP, nF})}_{\text{hyperbolic-secant flow }(-)} & \cdots & \underbrace{(\text{DP : UP, nF})}_{\text{hyperbolic flow }(-)} & \underbrace{(\text{UP : DP, nF})}_{\text{hyperbolic-secant flow }(-)} \\ \underbrace{(\text{DP : UP, nF})}_{\text{hyperbolic flow }(-)} & \cdots & \underbrace{(\text{UP : DP, nF})}_{\text{hyperbolic-secant flow }(-)} & \underbrace{(\text{DP : UP, nF})}_{\text{hyperbolic flow }(-)} \\ \vdots & \vdots & \vdots & \vdots \\ \underbrace{(\text{DP : UP, nF})}_{\text{hyperbolic flow }(-)} & \cdots & \underbrace{(\text{UP : DP, nF})}_{\text{hyperbolic-secant flow }(-)} & \underbrace{(\text{DP : UP, nF})}_{\text{hyperbolic flow }(-)} \end{bmatrix}_{(2n_1)\times(2m_1+1)} \tag{3.29}$$

- *For* $a_{110} < 0$ and $a_{220} < 0$,

$$\bigcup_{\alpha=1}^{2m_1+1}\bigcup_{\beta=1}^{2n_1}\underbrace{(a_{11\alpha},a_{12\beta})}_{(\text{XX, nF})} = \begin{bmatrix} \underbrace{(\text{DP : UP, nF})}_{\text{hyperbolic flow }(-)} & \cdots & \underbrace{(\text{UP : DP, nF})}_{\text{hyperbolic-secant flow }(-)} & \underbrace{(\text{DP : UP, nF})}_{\text{hyperbolic flow }(-)} \\ \underbrace{(\text{UP : DP, nF})}_{\text{hyperbolic-secant flow }(-)} & \cdots & \underbrace{(\text{DP : UP, nF})}_{\text{hyperbolic flow }(-)} & \underbrace{(\text{UP : DP, nF})}_{\text{hyperbolic-secant flow }(-)} \\ \vdots & \vdots & \vdots & \vdots \\ \underbrace{(\text{UP : DP, nF})}_{\text{hyperbolic-secant flow }(-)} & \cdots & \underbrace{(\text{DP : UP, nF})}_{\text{hyperbolic flow }(-)} & \underbrace{(\text{UP : DP, nF})}_{\text{hyperbolic-secant flow }(-)} \end{bmatrix}_{(2n_1)\times(2m_1+1)} \tag{3.30}$$

The lines of $x_1^* = a_{11\alpha}$ ($\alpha \in \{1, 2, \ldots, 2m_1 + 1\}$) are directrix flows.

3.4 [(2m_1), (2n_1)]-Inflection-Saddle Flows

Consider a polynomial dynamical system as

$$\begin{aligned}&\dot{x}_2 = a_{220},\\&\dot{x}_1 = a_{110}(x_1 - a_{111})^{2m_1}(x_2 - a_{121})^{2n_1}.\end{aligned} \tag{3.31}$$

The first integral manifold is

$$\begin{aligned}&-\frac{1}{2m_1-1}a_{220}\left[\frac{1}{(x_1-a_{111})^{2m_1-1}}-\frac{1}{(x_{10}-a_{111})^{2m_1-1}}\right]\\&=\frac{1}{2n_1+1}a_{110}\left[(x_2-a_{121})^{2n_1+1}-(x_2-a_{121})^{2n_1+1}\right]\\&\text{for } 2m_1-1\neq 0.\end{aligned} \tag{3.32}$$

For $(x_1^*, x_2^*) = (a_{111}, a_{121})$ with $\dot{x}_1 = 0$, the $[(2m_1+1), (2n_1)]$-order singularity exists with $\dot{x}_2 = a_{220} \neq 0$. For $(x_1^*, x_2^*) = (a_{111}, a_{121})$, the 1-dimensional flow direction is determined by $\dot{x}_2 \neq 0$. Thus,

$$\begin{aligned}&(a_{111}, a_{121}) = \underbrace{([(2m_1),(2n_1)]_{\text{II:II}}\text{US,pF})}_{[(2m_1),(2n_1)]\text{-increasing-inflection upper-saddle flow }(+)},\\&\text{for } a_{110}>0,\ a_{220}>0;\\&(a_{111}, a_{121}) = \underbrace{([(2m_1),(2n_1)]_{\text{DI:DI}}\text{LS,pF})}_{[(2m_1),(2n_1)]\text{-decreasing-inflection lower-saddle flow }(+)},\\&\text{for } a_{110}<0,\ a_{220}>0;\\&(a_{111}, a_{121}) = \underbrace{([(2m_1),(2n_1)]_{\text{DI:DI}}\text{US,nF})}_{[(2m_1),(2n_1)]\text{-decreasing-inflection upper-saddle flow }(-)},\\&\text{for } a_{110}>0,\ a_{220}<0;\\&(a_{111}, a_{121}) = \underbrace{([(2m_1),(2n_1)]_{\text{II:II}}\text{LS,nF})}_{[(2m_1),(2n_1)]\text{-increasing-inflection lower-saddle flow }(-)},\\&\text{for } a_{110}<0,\ a_{220}<0.\end{aligned} \tag{3.33}$$

The $([(2m_1), (2n_1)]_{\text{II:II}}\text{US,pF})$ is for a positive increasing-inflection upper-saddle flow. The $([(2m_1), (2n_1)]_{\text{DI:DI}}\text{LS,pF})$ is for a positive decreasing-inflection lower-saddle flow. The $([(2m_1), (2n_1)]_{\text{DI:DI}}\text{US,nF})$ is for a negative decreasing-inflection upper-saddle flow. The $([(2m_1), (2n_1)]_{\text{II:II}}\text{LS,nF})$ is for a negative increasing-inflection lower-saddle flow. The acronyms "II" and "DI" are for increasing and decreasing-inflection flows, respectively. US is for a upper-saddle flow, and LS is for a lower-saddle flow. The singular hyperbolic and hyperbolic-secant flows are illustrated in Fig. 3.7. Through such singular flows, the corresponding global bifurcations generate from a source (sink) flow to a network of hyperbolic and hyperbolic-secant flows.

From the above bifurcations, there is a non-singular system with $(2m_1) \times (2n_1)$-hyperbolic and hyperbolic-secant flows as

$$\begin{aligned}\dot{x}_1 &= a_{110} \prod_{s=1}^{2m_1} (x_1 - a_{11s}) \prod_{l=1}^{2n_1} (x_2 - a_{12l}),\\ \dot{x}_2 &= a_{220}.\end{aligned} \tag{3.34}$$

The first integral manifold is given by (Fig. 3.8)

$$\begin{aligned}&a_{j_2j_20} \sum_{\alpha_1=1}^{2m_1} \frac{1}{\prod_{\alpha_2=1,\alpha_2\neq\alpha_1}^{2m_1} (a_{j_1j_1\alpha_1} - a_{j_1j_1\alpha_2})} \ln \frac{|x_{j_1} - a_{j_1j_1\alpha_1}|}{|x_{j_10} - a_{j_1j_1\alpha_1}|}\\ &= a_{j_1j_10}\Big\{\frac{1}{2n_1+1}\big[(x_{j_2} - a_{j_1j_2\beta_1})^{2n_1+1} - (x_{j_20} - a_{j_1j_2\beta_1})^{2n_1+1}\big]\\ &+ \sum_{k=1}^{2n_1-1} \frac{1}{2n_1-k+1} b_{j_1j_2k}\big[(x_{j_2} - a_{j_1j_2\beta_1})^{2n_1-k+1} - (x_{j_20} - a_{j_1j_2\beta_1})^{2n_1-k+1}\big]\Big\}.\end{aligned} \tag{3.35}$$

where

$$\begin{aligned}b_{j_1j_21} &= \sum_{\beta_2=1,\beta_2\neq\beta_1}^{2n_1} (a_{j_1j_2\beta_1} - a_{j_1j_2\beta_2}),\\ b_{j_1j_22} &= \sum_{\substack{\beta_2,\beta_3=1;\beta_2,\beta_3\neq\beta_1\\(\beta_2<\beta_3)}}^{2n_1} \prod_{r=2}^{3} (a_{j_1j_2\beta_1} - a_{j_1j_2\beta_r}), \ldots,\\ b_{j_1j_2k} &= \sum_{\substack{\beta_2,\beta_3,\ldots,\beta_{k+1}=1;\\ \beta_2,\beta_3,\ldots,\beta_{k+1}\neq\beta_1\\(\beta_2<\beta_3<\cdots<\beta_{k+1})}}^{2n_1} \prod_{r=2}^{k+1} (a_{j_1j_2\beta_1} - a_{j_1j_2\beta_r}), \ldots,\\ b_{j_1j_2(2n_1-1)} &= \prod_{\beta_1=1,\beta_2\neq s_1}^{2n_1} (a_{j_1j_2\beta_1} - a_{j_1j_2\beta_2}).\end{aligned} \tag{3.36}$$

Phase portraits for the 2-dimensional systems with the hyperbolic and hyperbolic-secant flows at $(x_{j_1}^*, x_{j_2}^*) = (a_{j_1j_1\alpha}, a_{j_1j_2\beta})$ ($\alpha \in \{1, 2, \ldots, 2m_1\}, \beta \in \{1, 2, \ldots, 2n_1\}$) for the x_{j_1}-direction are presented in Fig. 3.6a–d for ($a_{110} > 0, a_{220} > 0$), ($a_{110} < 0, a_{220} > 0$), ($a_{110} > 0, a_{220} < 0$) and ($a_{110} < 0, a_{220} < 0$). The equilibriums of $(x_{j_1}^*, x_{j_2}^*) = (a_{j_1j_1\alpha}, a_{j_1j_2\beta})$ for the x_{j_1}-direction are for hyperbolic and hyperbolic-secant flow network matrices, which are generated from a two-dimensional flow to the corresponding network.

- *For* $a_{110} > 0$ and $a_{220} > 0$,

$$\bigcup_{\alpha=1}^{2m_1}\bigcup_{\beta=1}^{2n_1}\underbrace{(a_{11\alpha}, a_{12\beta})}_{\text{(XX, pF)}} = \begin{bmatrix} \underbrace{\text{(UP : DP, pF)}}_{\text{hyperbolic-secant flow }(+)} & \cdots & \underbrace{\text{(UP : DP, pF)}}_{\text{hyperbolic-secant flow }(+)} & \underbrace{\text{(DP : UP, pF)}}_{\text{hyperbolic flow }(+)} \\ \underbrace{\text{(DP : UP, pF)}}_{\text{hyperbolic flow }(+)} & \cdots & \underbrace{\text{(DP : UP, pF)}}_{\text{hyperbolic flow }(+)} & \underbrace{\text{(UP : DP, pF)}}_{\text{hyperbolic-secant flow }(+)} \\ \vdots & \vdots & \vdots & \vdots \\ \underbrace{\text{(DP : UP, pF)}}_{\text{hyperbolic flow }(+)} & \cdots & \underbrace{\text{(DP : UP, pF)}}_{\text{hyperbolic flow }(+)} & \underbrace{\text{(UP : DP, pF)}}_{\text{hyperbolic-secant flow }(+)} \end{bmatrix}_{(2n_1)\times(2m_1)} \tag{3.37}$$

- *For* $a_{110} < 0$ and $a_{220} > 0$,

$$\bigcup_{\alpha=1}^{2m_1}\bigcup_{\beta=1}^{2n_1}\underbrace{(a_{11\alpha}, a_{12\beta})}_{\text{(XX, pF)}} = \begin{bmatrix} \underbrace{\text{(DP : UP, pF)}}_{\text{hyperbolic flow }(+)} & \cdots & \underbrace{\text{(DP : UP, pF)}}_{\text{hyperbolic flow }(+)} & \underbrace{\text{(UP : DP, pF)}}_{\text{hyperbolic-secant flow }(+)} \\ \underbrace{\text{(UP : DP, pF)}}_{\text{hyperbolic-secant flow }(+)} & \cdots & \underbrace{\text{(UP : DP, pF)}}_{\text{hyperbolic-secant flow }(+)} & \underbrace{\text{(DP : UP, pF)}}_{\text{hyperbolic flow }(+)} \\ \vdots & \vdots & \vdots & \vdots \\ \underbrace{\text{(UP : DP, pF)}}_{\text{hyperbolic-secant flow }(+)} & \cdots & \underbrace{\text{(UP : DP, pF)}}_{\text{hyperbolic-secant flow }(+)} & \underbrace{\text{(DP : UP, pF)}}_{\text{hyperbolic flow }(+)} \end{bmatrix}_{(2n_1)\times(2m_1)} \tag{3.38}$$

- *For* $a_{110} > 0$ and $a_{220} < 0$,

$$
\bigcup_{\alpha=1}^{2m_1}\bigcup_{\beta=1}^{2n_1}\underbrace{(a_{11\alpha},a_{12\beta})}_{\text{(XX, nF)}}
=\begin{bmatrix}
\underbrace{\text{(DP : UP, nF)}}_{\text{hyperbolic flow }(-)} & \cdots & \underbrace{\text{(DP : UP, nF)}}_{\text{hyperbolic flow }(-)} & \underbrace{\text{(UP : DP, nF)}}_{\text{hyperbolic-secant flow }(-)}\\
\underbrace{\text{(UP : DP, nF)}}_{\text{hyperbolic-secant flow }(-)} & \cdots & \underbrace{\text{(UP : DP, nF)}}_{\text{hyperbolic-secant flow }(-)} & \underbrace{\text{(DP : UP, nF)}}_{\text{hyperbolic flow }(-)}\\
\vdots & \vdots & \vdots & \vdots\\
\underbrace{\text{(UP : DP, nF)}}_{\text{hyperbolic-secant flow }(-)} & \cdots & \underbrace{\text{(UP : DP, nF)}}_{\text{hyperbolic-secant flow }(-)} & \underbrace{\text{(DP : UP, nF)}}_{\text{hyperbolic flow }(-)}
\end{bmatrix}_{(2n_1)\times(2m_1)}
\tag{3.39}
$$

- *For* $a_{110} < 0$ and $a_{220} < 0$,

$$
\bigcup_{\alpha=1}^{2m_1}\bigcup_{\beta=1}^{2n_1}\underbrace{(a_{11\alpha},a_{12\beta})}_{\text{(XX, nF)}}
=\begin{bmatrix}
\underbrace{\text{(UP : DP, nF)}}_{\text{hyperbolic-secant flow }(-)} & \cdots & \underbrace{\text{(UP : DP, nF)}}_{\text{hyperbolic-secant flow }(-)} & \underbrace{\text{(DP : UP, nF)}}_{\text{hyperbolic flow }(-)}\\
\underbrace{\text{(DP : UP, nF)}}_{\text{hyperbolic flow }(-)} & \cdots & \underbrace{\text{(DP : UP, nF)}}_{\text{hyperbolic flow }(-)} & \underbrace{\text{(UP : DP, nF)}}_{\text{hyperbolic-secant flow }(-)}\\
\vdots & \vdots & \vdots & \vdots\\
\underbrace{\text{(DP : UP, nF)}}_{\text{hyperbolic flow }(-)} & \cdots & \underbrace{\text{(DP : UP, nF)}}_{\text{hyperbolic flow }(-)} & \underbrace{\text{(UP : DP, nF)}}_{\text{hyperbolic-secant flow }(-)}
\end{bmatrix}_{(2n_1)\times(2m_1)}
\tag{3.40}
$$

The lines of $x_1^* = a_{11\alpha}$ ($\alpha \in \{1, 2, \ldots, 2m_1\}$) are directrix flows.

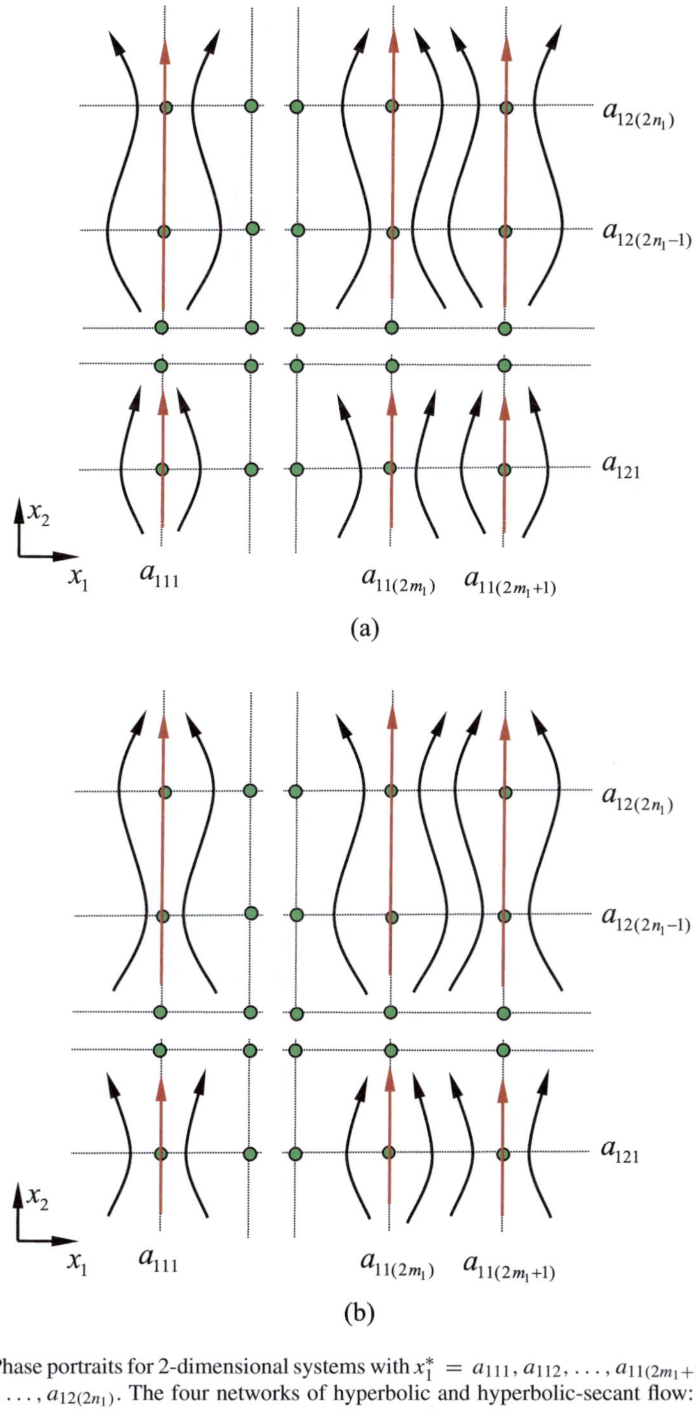

Fig. 3.6 Phase portraits for 2-dimensional systems with $x_1^* = a_{111}, a_{112}, \ldots, a_{11(2m_1+1)}$ and $x_2^* = a_{121}, a_{122}, \ldots, a_{12(2n_1)}$. The four networks of hyperbolic and hyperbolic-secant flow: **a** ($a_{110} > 0, a_{220} > 0$), **b** ($a_{110} < 0, a_{220} > 0$), **c** ($a_{110} > 0, a_{220} < 0$), **d** ($a_{110} < 0, a_{220} < 0$)

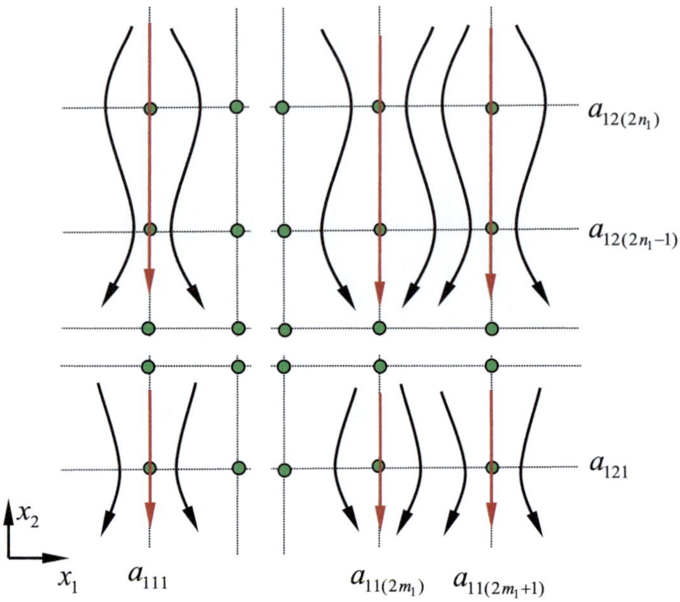

(c)

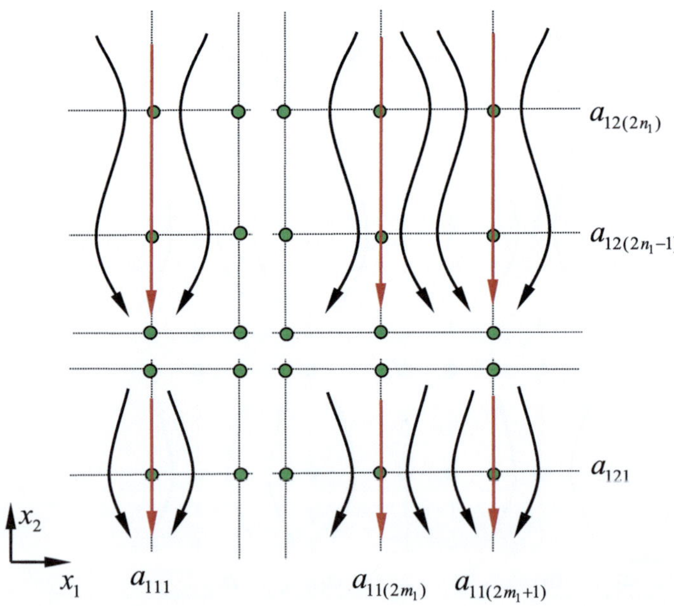

(d)

Fig. 3.6 (continued)

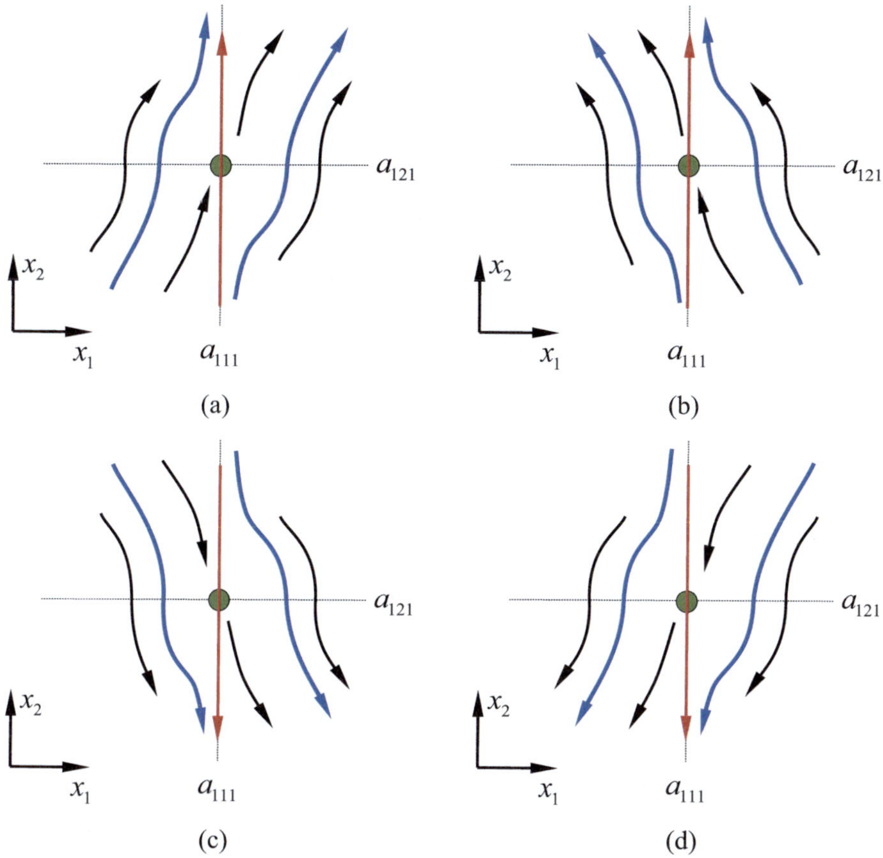

Fig. 3.7 Phase portraits of the ([2m_1], [2n_1])-increasing and decreasing-inflection saddle singular flows of $(x_1^*, x_2^*) = (a_{111}, a_{121})$ in the x_1-direction and constant speed in the x_2-direction for 2-dimensional polynomial systems: **a** positive singular increasing-inflection upper-saddle flow ($a_{110} > 0, a_{220} > 0$), **b** positive singular decreasing-inflection lower-saddle flow ($a_{110} < 0, a_{220} > 0$), **c** negative singular decreasing-inflection upper-saddle flow ($a_{110} > 0, a_{220} < 0$), **d** negative singular increasing-inflection lower-saddle flow ($a_{110} < 0, a_{220} < 0$)

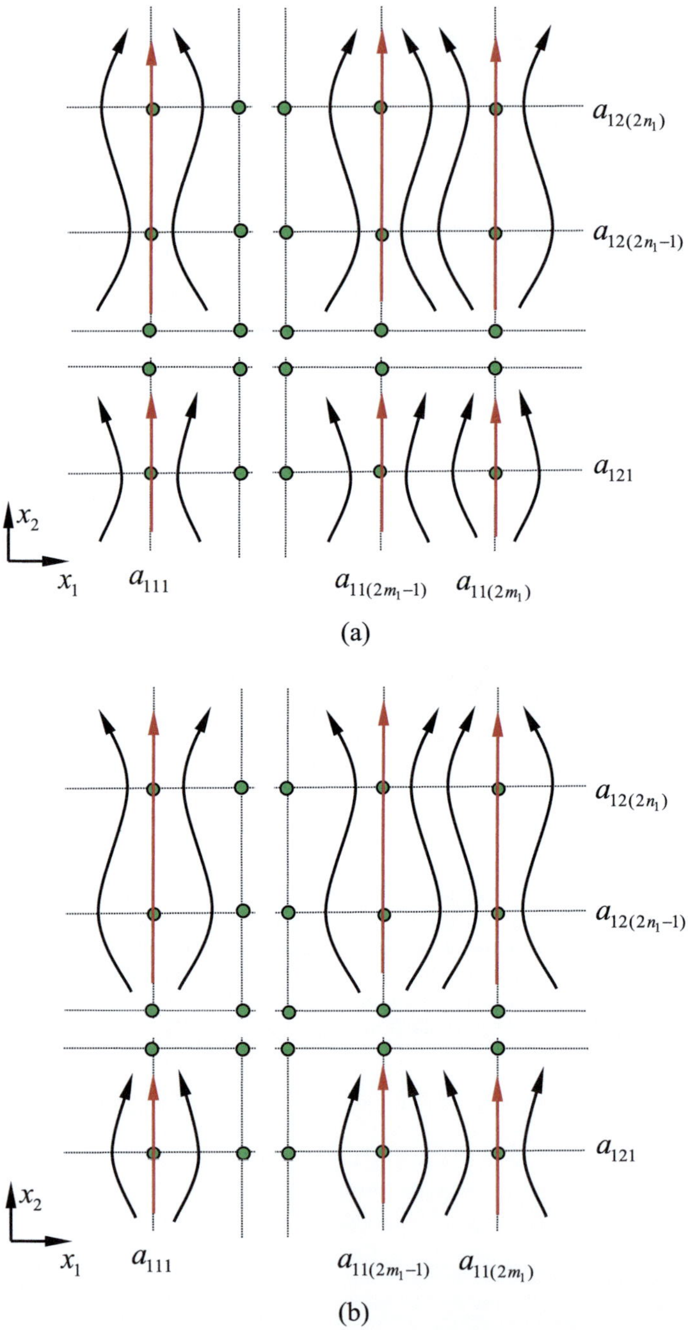

Fig. 3.8 Phase portraits for 2-dimensional systems with $x_1^* = a_{111}, a_{112}, \ldots, a_{11(2m_1)}$ and $x_2^* = a_{121}, a_{122}, \ldots, a_{12(2n_1)}$. The four networks of hyperbolic and hyperbolic-secant flow: **a** ($a_{110} > 0, a_{220} > 0$), **b** ($a_{110} < 0, a_{220} > 0$), **c** ($a_{110} > 0, a_{220} < 0$), **d** ($a_{110} < 0, a_{220} < 0$)

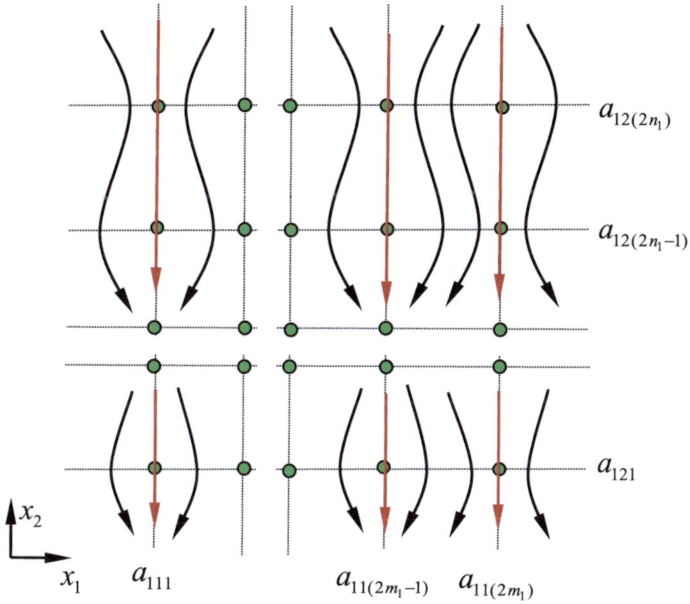

(c)

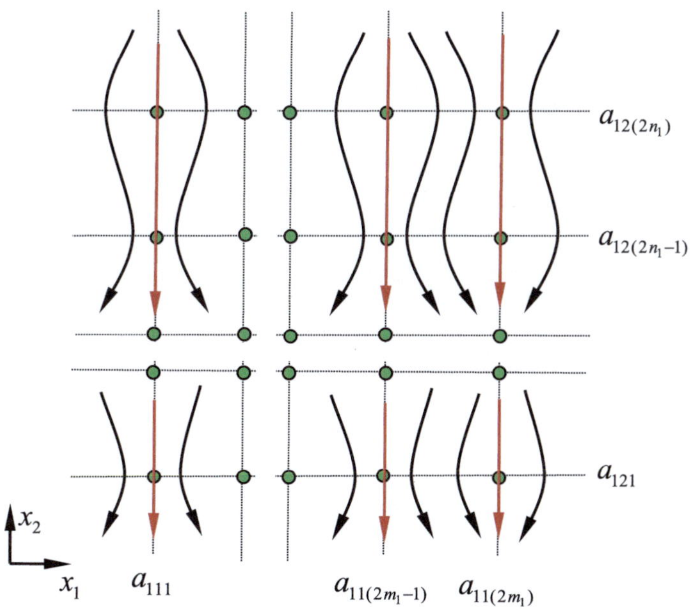

(d)

Fig. 3.8 (continued)

Correction to: Two-Dimensional Constant and Product Polynomial Systems

Correction to:
A. C. J. Luo, *Two-dimensional Constant and Product Polynomial Systems*,
https://doi.org/10.1007/978-981-96-5515-1

The original version of this book has been updated with the equations. The book and the chapter have been updated with the changes.

The updated version of this book can be found at
https://doi.org/10.1007/978-981-96-5515-1

A. C. J. Luo, *Two-dimensional Constant and Product Polynomial Systems*,
https://doi.org/10.1007/978-981-96-5515-1_4

Index

A. C. J. Luo, *Two-dimensional Constant and Product Polynomial Systems*,
https://doi.org/10.1007/978-981-96-5515-1

O

P